William Miles Maskell

An Account of the Insects Noxious to Agriculture and Plants in New Zealand

William Miles Maskell

An Account of the Insects Noxious to Agriculture and Plants in New Zealand

ISBN/EAN: 9783744649889

Printed in Europe, USA, Canada, Australia, Japan

Cover: Foto ©berggeist007 / pixelio.de

More available books at **www.hansebooks.com**

AN ACCOUNT

OF THE

INSECTS NOXIOUS TO AGRICULTURE AND PLANTS

IN

NEW ZEALAND.

THE SCALE-INSECTS
(COCCIDIDÆ).

By W. M. MASKELL, F.R.M.S.,
REGISTRAR OF THE UNIVERSITY OF NEW ZEALAND.

WELLINGTON:
BY AUTHORITY: GEO. DIDSBURY, GOVERNMENT PRINTER.
1887.

MR. MASKELL'S Account of the Scale-Insects occurring in New Zealand is published by the State Forests and Agricultural Department, under the instructions of the Hon. John Ballance, Commissioner of State Forests.

Wellington, 31st March, 1887.

CONTENTS.

Chapter.	Page
Glossary of Terms and Phrases	1
I. Introductory	5
II. Characters, Life-history, and Metamorphoses of Coccididæ	8
III. Products of the Coccididæ (Honeydew; Black Fungus)	14
IV. Checks to Increase of Coccididæ, Parasites, etc.	18
V. Remedies against Coccididæ	24
VI. Catalogue of Insects and Diagnosis of Species	37
Groups—	
Diaspidinæ	39
Lecanidinæ	62
Hemicoccidinæ	87
Coccidinæ	88
Index of Plants and the Coccididæ attacking each	111
Index of Genera and Species	115

PREFACE.

THE number and variety of the insect pests which live on the plants of New Zealand, whether native or introduced, and the damage which they frequently do, form the excuse for the appearance of this work. The descriptions of these insects in the Transactions of the New Zealand Institute, or in works published in Europe and America, are not easily accessible to the general reader, and are also much scattered and fragmentary. It was thought therefore that the time had arrived when the information which might be useful to gardeners and tree-growers, as well as to students, might be summarized and brought together in a compendious form, and the present volume is an attempt towards this.

In order to render this work complete a second volume is necessary, which should include the large number of other destructive insects preying upon various plants. For example, the "pine-blight" (*Kermaphis*), the "American blight" (*Eriosoma*), the "black leech" (*Tenthredo*), the cabbage caterpillar, the turnip "fly," the various aphides on roses, geraniums. &c., the grass-grub (*Odontria*), the codlin-moth, the borers, weevils, wireworms, and a number of others are in different places damaging trees and plants, and it would be useful to collect in one volume information regarding them. The author has had in contemplation the preparation of such a volume, and it is hoped that it may be at some future time published.

Meanwhile the present is offered as, at least as far as it goes, a full description of one of the most general as well as the most noxious families of plant-parasites. The plates have been especially prepared with a double object: first, that gardeners and tree-growers might be able easily to recognize the kind of insect which might happen to be damaging their plants; and, secondly, that the student who should desire to know more of this curious family might have enough details indicated to guide him in his investigation. For the first purpose the figures have been coloured as near to nature as possible: for the second a few anatomical details have been introduced. The printing of these plates has been executed by Mr. Potts, lithographer to Mr. A. Willis, of Wanganui, and it is hoped that the reader may be well satisfied with the care and trouble which have been bestowed upon them.

The author is sensible that this volume may contain numerous imperfections; but these will not, he trusts, be attributed to culpable ignorance or carelessness.

EXPLANATION OF TERMS USED IN THE FOLLOWING PAGES.

Abdomen. The posterior half of the body of male or female, whether joined to the anterior half or slightly separated, segmented or not.

Abdominal cleft. A narrow slit in the extremity of the abdomen of *Lecanidinæ* and the full-grown *Hemicoccidinæ* only. (Plate ix., Fig. 1, *b*, *c*.) On the upper side of the body are seen the

Abdominal lobes, two minute, divergent, triangular or conical, excrescences, one on each side of the cleft, in *Lecanidinæ*, usually bearing one or more hairs. (Plate xi., Fig. 3, *b, c.*)

Abdominal spike. A more or less long, tubular or semi-tubular, pointed process terminating the abdomen of the male in all species, and serving as a sheath for the penis, which is a long, white, soft tube with recurved hairs. (Plate ii., Fig. 3 ; Plate xxi., Fig. 1, *k*.)

Anal ring, anogenital ring. An orifice situated near the abdominal extremity of the female, either simple or compound, hairless or bearing several hairs. (Plate ii., Fig. 1.)

Anal tubercles. Exhibited only by the *Coccidinæ* and by the larvæ of *Hemicoccidinæ*: two more or less conspicuous projecting processes at the abdominal extremity of the female, without any cleft, and in most instances projecting beyond the edge ; usually bearing setæ. (Plate ii., Fig. 2, *c, d.*)

Antennæ. Two jointed organs ("feelers") projecting from the anterior portion of the body, of variable length. (Plate i., Figs. 9, 10, 11, types.)

Apodous. Without feet.

Apterous. Without wings.

Bucca, buccal. The mouth ; belonging to the mouth.

Carina, carinated. A keel or raised ridge ; keeled.

Cephalic region. That part of the insect, male or female, which bears the eyes, antennæ, and mouth, but not including the first pair of feet.

Clavate. Club-shaped ; somewhat knobbed.

Claw. The hooked terminating joint of the foot. (Plate i., Fig. 6, *cl.,* type.)

Coxa. The first joint of the foot, springing directly from the underside of the thoracic region. (Plate i., Figs. 6 *c*, 7 *c*.)

Digitules. Appendages observed on the feet, and often useful for distinguishing species. Usually there are two pairs. The "upper pair" spring from the upper side of the extremity of the tarsus, and are generally long, fine hairs, terminating in a knob. The "lower pair" spring from the base of the claw, and are usually broader and more trumpet-shaped than the upper ones. (Plate i., Fig. 8, type.) Sometimes either pair, or both, may be absent. In *Cælostoma wairoense* there are no "upper" digitules, and 24 "lower" ones on the foot of the male. (Plate xxi.)

EXPLANATION OF TERMS.

Dimerous. Two-jointed.

Dorsum. The upper side of the body when the insect is in its natural position.

Dorsal. On the upper side or dorsum.

Eyes. Two coloured, granular or simple, round organs on the cephalic region of the female, near the base of the antennæ (Plate xiv., Fig. 2, *k.*; Plate xx., *b*); two, or four, coloured, granular, simple or facetted, on the head of the male (Plate i., Figs. 14, 15; Plate xxi., Fig. 2, *b*).

Femur. The second joint of the feet, next the coxa, joined to it by the false joint "trochanter." (Plate i., Figs. 6 *f*, 7 *f*.)

Fringe. A portion of the excreted substance, cotton or wax, produced by the spinnerets on the edge of the body in certain *Lecanidinæ* and *Hemicoccidinæ*. It may be in the form of long glassy threads (*Planchonia*) or of more or less broad flat plates (*Ctenochiton*). (Plate vii., Figs. 2 *d*, 3 *a*; Plate xii., Fig. 2, *a*, *b*, *c*.)

Halters. A minute organ, situated just behind the wings of the males, and of which the use, either in this family or in the Diptera, has not been satisfactorily ascertained. It is often termed the "balancer." In the house-fly it has been thought to represent an organ of hearing. In Coccids it is furnished with one or more hooked bristles, and Mr. Comstock affirms that these are, probably for some purposes of flight, hooked into the posterior edges of the wings.* (Plate i., Fig. 17; Plate xxi., Fig. 1, *m.*)

Honeydew. A substance of a glutinous character produced by many species, and falling in spray from them on the leaves. (See Chap. III.)

Larva. The first stage in the insect's life after emerging from the egg.

Lobes, in the *Diaspidinæ,* are minute, flat, more or less rounded projections, two or more, seen on the edge of the abdominal extremity, usually interspersed with spines and hairs (Plate iii., Figs. 1, 3, 4, 5, *l*); in the *Lecanidinæ*, are two triangular or conical projections, usually bearing hairs, on the dorsal side of the body, one on each side of the abdominal cleft (Plate xi., Fig. 3, *b*, *c*).

Mentum. A kind of secondary rostrum, or "under-lip," not altogether tubular, but rather a deepish trough, through which the rostral setæ pass after leaving the rostrum. It may have one, two, or three joints. It is not noticeable in the Diaspidinæ. (Plate i., Fig. 5, *b.*)

Metamorphosis. A change of form. For the number and characters of these see Chap. II.

Moniliform. Like a string of beads.

Monomerous. With a single joint.

Multilocular. With several divisions: a term applied to the spinneret orifices of some insects, distinguishing them from "simple" orifices, which show only a single tube. Multilocular orifices exhibit a bundle of tubes enclosed together. (Plate i., Fig. 4, *c*, *d*, *p*; Plate xviii., Fig. 2, *e.*)

Nervure. A strong vein which, starting from the attachment of the wing of the male, runs along the anterior edge of the wing, a little within it: at about half its length a branch runs obliquely towards the posterior edge. (Plate i., 16; Plate xix., *f*; Plate xxi.)

Normal. According to rule—not exceptional.

Ocelli. Two, four, or six minute circular simple organs, on the head of the male: probably organs of vision. In the *Monophlebidæ* they would seem

---

* Report of the Entomologist, U.S. Dep. of Agric. 1880, p. 277, note.

EXPLANATION OF TERMS.

to be replaced by a smooth rounded protuberance behind the eye. (Plate i., Fig. 14, *oc.*; Plate viii., Fig. 2, *k, m;* Plate xxi., Fig. 2, *b.*)

Ovisac. The cottony bag or nest formed by certain species of *Lecanidinæ* and *Coccidinæ* for the reception of their eggs. (Plate xii., Fig. 1, *a, b, c;* Plate xix., *a, b, c.*)

Peduncle, pedunculated. A stalk; stalked.

Pellicle. The skin of an earlier stage, cast off at each metamorphosis; used by the *Diaspidinæ* and by one genus of *Lecanidinæ* in the formation of the puparium or test. (Plate i., Fig. 3, *a, b,* Plate vii., Fig. 2, *b.*)

Polymerous. Many-jointed.

pa. The last stage of the male insect before emerging winged.

Puparium. The shield, covering, or "scale" of the *Diaspidinæ.* (Plate i., Fig. 3, *c;* Plates iv., v., vi.)

Rostral setæ. Three or, in a few cases, four long, fine, curling, tubular bristles springing from the rostrum, and often passing through a mentum; used for insertion into the tissues of a plant and sucking their contents. (Plate i., Fig. 5; Plate vi., only one being here shown, from the smallness of the drawings.)

Rostrum. A more or less conical, tubular, projecting organ, or beak, protruding from the underside of the cephalic region, or between the first pair of feet. It is absent in the adult female *Cœlostoma.* It is the "mouth" of the insect. (Plate i., Fig. 5; Plate iv., Fig. 5.)

Sac. The cottony, bag-like covering or nest produced by the spinnerets and concealing the insect in many of the *Coccidinæ* and some *Lecanidinæ.* (Plate xv., Fig. 1, *c;* Fig. 2, *b.*)

Scale. The shield or puparium of the *Diaspidinæ.* The word is commonly used to designate the outward appearance of insects of the whole family, which are indiscriminately called "scale-insects," although many of them form no shield whatever.

Secretion may be of various kinds. It is matter produced by internal organs, and expelled through the "spinnerets." In the *Diaspidinæ* the secreted portion of the puparium (that is, all except the pellicles) is made up of fine, closely-woven fibres, forming the "scale." In the *Lecanidinæ* it probably exudes originally as fine fibres, but these become agglomerated in some cases in a waxy or horny mass, or in others are loosely collected as cotton. In the *Coccidinæ* the secretion is usually cottony, or powdery like meal. *Cœlostoma* secretes all three—wax, cotton, and meal. In some instances, as in *Carteria lacca,* of Africa, the wax, called "shellac," is abundant enough to be commercially valuable; or, as in the Chinese *Ericerus Pe-la* it can be used for making candles.

Seta. A bristle—a long stiff hair.

Setose. Bearing a few bristles.

Spinnerets. Organs observed in various parts of the body, producing the waxy, cottony, or mealy matter. They consist of cylindrical internal tubes, sometimes ending on the skin, sometimes protruding outside it in the form of tubes, spines, or conical hairs. In the former case the orifices show them to be in some instances simple, and in others compound tubes.[*] In the *Diaspidinæ,* besides being scattered over the body, the spinnerets are arranged in groups on the last abdominal segment, and these groups afford excellent characters

---

[*] Minute anatomical details are unsuitable for this work. The student may consult Targioni-Tozzetti, "Studie sulle Cocciniglie," cap. ii., p. 26.

for specific distinctions. (Plate i., Fig. 4, for types of various spinnerets; Plate iii., groups of spinnerets of *Diaspidinæ*.)

Spiracles. "Breathing organs:" the orifices in the body of the tracheæ or tubes conveying air to the blood. In the *Lecanidinæ* they are usually four; simple circles, near the edge of the body, and with a few strong spiny hairs near them. In the *Coccidinæ* they are often numerous. (Plate ii., Fig. 4; Plate xx., *n.*)

Spiracular spines. Spiny hairs, usually three in number, of which one is rather long, close to the spiracles, in the *Lecanidinæ*.

Stigma, stigmatic spines. Terms sometimes employed for spiracles, &c.

Tarsus. The fourth joint of the feet, between the tibia and the claw. Its consisting of one joint (monomerous) is a distinctive character of the whole family. (Plate i., Fig. 6, *ta.*; Fig. 7, *ta.*)

Test. The waxy, glassy, or horny covering produced through the spinnerets and concealing the insect in many *Lecanidinæ* and some *Coccidinæ*. In this work it is not applied to the "scale" of *Diaspidinæ* or to cottony secretions.

Thoracic band. An appearance seen on the thoracic region in the male, looking like a broad transverse ribbon.

Thoracic region, thorax. That part of the female or the male which bears the three pairs of feet, when the feet are present; or, if the feet are absent, the middle portion of the body, segmented or not.

Tibia. The third joint of the feet, next the femur. (Plate i., Fig. 6, *ti.*; Fig. 7, *ti.*)

Tracheæ. Tubes ramifying throughout the body, conveying air to the blood. Their orifices are the spiracles. The tracheæ, as in other insects, appear as if constructed of a network of fine spiral wires. (Plate ii., Fig. 4 *d*; Plate xx., *n.*)

Trochanter. A small articulation, not a distinct joint (something like a knee-cap) of the feet, between the coxa and the femur. (Plate i., Fig. 6, *tr.*; Fig. 7, *tr.*)

Trimerous. Three-jointed.

Ventral. On the under-side, the insect being in its proper position.

NEW ZEALAND SCALE-INSECTS
(COCCIDIDÆ).

CHAPTER I.

INTRODUCTORY.

INSECTS are divided by naturalists into several principal orders, the distinguishing marks of which are generally very well defined—for example, the butterflies and moths belong to the order *Lepidoptera*, the dragon-flies to the *Neuroptera*, the common house-flies to the *Diptera*, and so on. These orders are founded upon the characters and arrangement of the wings. They are subdivided into families, and these again into genera and species. One of the orders is that of the HEMIPTERA, which is composed of the two following sections :—

HEMIPTERA-HETEROPTERA, including the bugs, water-beetles, &c.

HEMIPTERA-HOMOPTERA, including the crickets, cuckoo-spits, plant-lice (Aphides), leaf-hoppers (Psyllids), scale-insects (Coccids), &c.

The insects treated of in this volume are therefore placed as follows :—

    Class—INSECTA.
    Order—HEMIPTERA.
    Section—HOMOPTERA.
    Family—COCCIDIDÆ.

The genera and species will be found in their places.

The common English name for this family—" scale-insects " —is not very appropriate. Some few of them have the appearance of small thin scales on leaves or twigs, but many have not. Nor are the German appellations—" gall-insekten " or " schild-lause " —more appropriate. Gardeners have given to some of them the name of "mealy-bug," which, although decidedly neither elegant nor euphonious, very fairly represents the character of that particular portion of the family.

The origin of the name "Coccididae," or, as abbreviated often in this volume, "Coccids," is found in the old Greek word "κοκκος," denoting a rich red dye, which was much admired by the Greeks and Romans, and which was procured from the insect now known as *Kermes vermilio* (the *Coccus ilicis* of Linnæus). When the cochineal insect was discovered in Mexico it soon overpowered all the others, producing commercial dyes, and from it has come the title "Coccid," now applied to the whole family. Cochineal itself has of late years been pushed aside to a great extent by the aniline (coal-tar) dyes; yet it is still used for many purposes. This insect lives on the leaves of cactus. Amongst the New Zealand species described in this work will be found one, *Dactylopius alpinus*, which produces a red dye similar to, though probably not equal to, cochineal. Before the discovery of aniline dyes it might possibly have been worth while to cultivate this insect for its dye; but this would scarcely answer now.

The Coccididae are, in some parts of the world, very injurious to vegetation. They seem to affect principally the warmer temperate regions. California, Florida, the Cape of Good Hope, the southern parts of Australia, Southern France and Northern Italy, and New Zealand are countries in which they are found out-of-doors in the greatest numbers. In England they are less troublesome in the open air, though in greenhouses and hot-houses they abound; but, in places under glass, every gardener ought to be able to get rid of them without difficulty. For its extent New Zealand seems to furnish a larger number than any other country. The humidity of its climate and the absence of anything like severe winters in most parts of it are quite congenial to Coccids; and there is scarcely a tree in its forests or in its gardens, whether native or introduced, which is not subject to their attacks.

It has not been thought necessary to include in this work a list of the books and essays written on this family of insects. The list would be a very long one; but, besides that many of the books would not be obtainable here, it would be found that very many authors have done nothing more than copy— often quite blindly and unintelligently—what others had said before them; moreover, most of them are out of date. The student or the horticulturist desiring to know more about Coccids not found in New Zealand may find full details in the reports of the Agricultural Department of the United States Government, in Dr. V. Signoret's "Essai sur les Cochenilles" (Paris), in papers by Miss Emily Smith (American naturalist, 1878–80), &c. The American Departmental Reports of Professors Riley and Comstock, Mr. Hubbard, and Mr. L. Howard contain most valuable information. English works on the subject are mostly fragmentary or inaccurate; but Mr. Douglas, of Lewisham, has lately begun to discuss the Coccids in England in a systematic manner, and probably before long others will follow suit. In India, Mr. T. W. Atkinson, of Calcutta, is studying the family.

Natural science in these days tends ever more and more towards specialization, and the boundaries of scientific classes, orders, families, &c., are becoming always more and more narrowed. The student can find his time quite sufficiently occupied nowadays in the thorough investigation of so (comparatively) small a portion of the animal kingdom as is presented by the Coccids of even only one country; and the present work may not be without value to future workers in this direction. To the farmer, the gardener, the fruit-grower, and the owner of pleasure-grounds it is believed that the following chapters will also supply information at the same time correct, intelligible, and useful.

CHAPTER II.

CHARACTERS, LIFE-HISTORY, AND METAMORPHOSES OF THE COCCIDIDÆ.

The first principal character separating the Coccididæ from the rest of the Homoptera, and distinguishable without microscopic examination, is the absence of wings in the females at all stages of their existence.

The second principal character is the absence of any apparatus for feeding and digesting in the males.

From these two characters it follows that the females can only extend their operations by, at the best, crawling from plant to plant, or by being carried about by birds or other agency; also that the males cannot enjoy more than a very short existence, their work being entirely confined to impregnating the females. Hence, in any endeavours to destroy these insects, the males may be disregarded, and the females only attended to.

Other distinguishing characters, chiefly microscopic, are—

1. The presence of only one joint in the tarsus or fourth joint of the leg, in both males and females (Plate i., Figs. 6 and 7, *la*);

2. The presence of only a single claw terminating the leg in both males and females (Plate i., Figs. 6 and 7);

3. The presence of only two wings, with two halteres, in the full-grown males (Plate xxi);

4. The presence of two or more eyes or ocular tubercles, in addition to the ordinary pair of eyes, in the full-grown males (Plate i., Fig. 14; Plate viii., Fig. 1, *k*, *m*).

I. The Female Insect.

In general outward appearance the female insects present very variable forms. They may be either naked, or covered over with some kind of a shield, which may be fibrous, or waxy, or cottony, or they may have simply a thin powdery meal scattered over them. The covered insects are, of course, stationary, although in some cases, before reaching their full development, they move about, carrying their houses with them. The naked insects may be either stationary or active.

They attach themselves either to the bark or stem of a plant or to the leaves. In the latter case it is rare to find them on the upper side; but, on turning over a leaf, the under-surface is frequently found covered thickly with them.

In many cases they exude, in the form of minute globules, a whitish, thick, gummy secretion, answering probably to the "honeydew" of the Aphididæ. This secretion drops from them on to the plant, and from it grows a black fungus, which soon gives an unsightly appearance to the plant. This fungus or "smut" is an almost invariable indication that a plant is attacked by insects,* and may, indeed, give a useful warning to tree-growers. It is not, however, produced in appreciable quantities by all species.

The manner of feeding upon the plant is the same as in all the families of Homoptera—namely, by means of a protruding rostrum, beak, or trunk, situated on the under-side of the insect. As there is not, in the female Coccididæ, any well-defined division between the head and the rest of the body, this rostrum is seen, on turning over the insect, in the form, usually, of a minute conical projection between, or nearly between, the first pair of legs, if the legs are present, or a little within the circumference, if the legs are absent (Plate i., Fig. 5). An ordinary lens will generally show, springing from the point of the conical rostrum, three or four longish, very fine, curling bristles. These bristles are, in fact, hollow tubes, and the insect, inserting them into the leaf or bark of the plant, sucks through them its food. It is thus plain that, with often great numbers of scale-insects sucking at it—pumping, as it were, its life-blood through their rostra—a plant must of necessity suffer greatly.

Birds do not, as a rule, seem to care much about eating the Coccididæ, whose work is thus little interfered with by them. The "white-eye" (Zosterops) or "blight-bird" has been seen feeding on scale-insects; but its visits are few and far between, and its assistance to the gardener in this respect not great. The Coccididæ are, however, much subject to attacks

---

* Not necessarily a Coccid insect: the fungus may also grow on the honeydew of Aphis; but it is easy to recognize the difference between these insects. In every case there is some insect at work where the fungus is.

from Hymenopterous parasites, of which some account will be found in a subsequent chapter (Chap. IV.).

The effects of the Coccididæ are not confined altogether to damage to plants: there are some species producing materials useful to man. For example, *Coccus cacti* produces cochineal; *Carteria lacca* produces shellac; *Ericerus pé-la* is used by the Chinese for candles: and others might be mentioned. But, so far, no New Zealand species appears to be of any commercial use. *Dactylopius alpinus* makes a rather rich red dye in alcohol; *Cœlostoma zealandicum* constructs thick, waxy coverings, which might possibly be turned to some account; but even these are probably not worth much.

Groups.

The groups into which the Coccididæ are, in this work, divided are as follow:—

1. Female insects constructing for themselves shields composed partly of secretion, partly of the pellicles discarded from earlier stages; abdomen not cleft; legs lost at full growth. DIASPIDINÆ.
2. Female insects naked, or covered with shields of secretion, either waxy, horny, cottony, or felted; abdomen in all stages cleft; legs either lost or retained at full growth. LECANIDINÆ.
3. Female insects naked, or covered with shields of waxy secretion; abdomen of larva ending in prominent processes, abdomen of adult cleft; legs either lost or retained at full growth. HEMICOCCIDINÆ.
4. Female insects naked, or covered with secretion either waxy, cottony, or felted; abdomen in all stages ending in prominent processes; legs either lost or retained at full growth. COCCIDINÆ.

Life-history.

The life-history of the insects in the above groups is as follows:—

All of them pass through four stages of existence: 1, the

egg; 2, the young larva; 3, the second stage of life, or "pupa;" 4, the adult, or full-grown insect.

1. *The egg.* This is, in all cases, of regularly-oval form, the colour varying from white to yellow or red (see Plate i., Fig. 1). It may be produced in great numbers, and in some cases several times in a year. As a general rule, the female ejects the eggs from her body; but there are some species, notably in the group *Lecanidinæ*, where the eggs are hatched within the body, the insect being thus, in a manner, viviparous.

2. *The young larva* (Plate i., Fig. 2). This is of precisely the same form both for the male and the female—or, rather, perhaps it should be said that no definite character has yet been discovered to show which are male and which are female larvæ. Neglecting slight variations of form, the larva is very minute—seldom more than about $\frac{1}{50}$in. long, often as small as $\frac{1}{100}$in.—oval, flattish, possessing a rostrum and accompanying bristles (setæ), six legs, and two antennæ: and in all species it is fairly active, travelling as soon as hatched over the plant in search of food.

3. *The second stage.* Here the first distinction is noticeable between the male and the female in most cases; but this distinction usually depends not so much upon the form of the insect as upon the character of the covering it makes for itself. Confining ourselves at present to the female, there are differences now noticeable between the groups. In the *Diaspidinæ* the insect begins by slipping out of the skin of the larva; but it does not cast it aside altogether: it makes use of the old skin as part of its covering. Adding to it a small portion of fibrous secretion—produced by organs called "spinnerets," which will be noticed presently—it attaches itself to the plant by its rostrum and setæ, and lies, inert and stationary, under a little shield composed half of its old skin and half of secretion. As it also, in entering this stage, loses its legs altogether, it must remain in the position it has chosen for the rest of its life. In the *Lecanidinæ* and in the *Coccidinæ* the skin of the larva is thrown away altogether, and the female in her second stage takes up a new position, in which she may be either naked or covered with a thin coat of secretion, active or stationary, retaining her legs in most cases, or losing them in some instances. In all the groups there is almost always some approach to the form of the full-grown insect noticeable in this second stage.

4. *The full-grown insect.* Here there is almost unlimited variety of form, colour, and habits. The insects may be naked or covered, active or stationary. In the *Diaspidinæ* the process just described is repeated: the female slips out of her second skin, but still keeps both it and the first over her, adding more fibrous secretion from the spinnerets; so that, in fact, she lies an inert, legless, slug-like object, under a covering composed partly of the two skins, partly of secretion. (See Plate i., Fig. 3 : *a* is the discarded larval skin, *b* the discarded skin of the second stage, both being used as part of the shield. In the genus *Aspidiotus* these skins would be in the centre instead of at one end.) In the *Lecanidinæ* (except in one single genus) and the *Coccidinæ* the second skin is discarded altogether; but the insect may either construct a new shield or remain naked, may be either with or without legs, either active or stationary. Once this last stage of her existence entered upon, the female prepares for laying her eggs. In most species the services of a male are needed; in some, as far as can be made out after investigation of many years, no males are found. The female, if naked, either hatches her eggs in her own body or lays them on the plant; if covered, she fills her shield with the eggs. The naked insects often cover the eggs themselves—e.g., *Lecanium hemisphæricum*; or, again, deposit them in an ovisac, a mass of cottony secretion—e.g., *Pulrinaria camellicola* or *Icerya purchasi*.

II. THE MALE INSECT.

It has been remarked above that, as the full-grown males of the *Coccididæ* are destitute of any organs for feeding whatsoever, there is no reason for making systematic attacks on them for economical purposes. Their function is simply to impregnate the females, and their life at this stage must necessarily be very brief. It will suffice in this place to observe that in all cases these males are small, two-winged flies, their size varying from about $\frac{1}{40}$ in. to $\frac{1}{4}$ in. in length; colour usually yellow or red; wings longer than the body, hyaline (glassy) and often iridescent, and, in repose, lying flat, partly crossing each other. The antennæ are long, slender, and hairy, consisting of nine or ten joints. The legs are also slender and hairy, the tarsus having only one joint, and terminating in a single claw. The insects are generally very active. Types of antenna, foot, wing and haltere, and a diagram of the arrangement of the eyes and ocelli, are given in Plate i., Figs. 7, 12, 13, 14, 15, 17.

The males are thus so small and rapid in their movements that it is difficult in most cases to find them in a free state. The usual way to procure them is by hatching them from the pupæ. In their course of life they pass through four stages, as do the females—viz.: 1, the egg; 2, the larva; 3, the pupa; 4, the full-grown insect.

1. *The egg* is, as far as can be made out, precisely the same as that of the female, though Dr. Signoret believes that in one or two species there may perhaps be minute differences.

2. *The larva* is, as stated above, similar to that of the female.

3. *The pupa.* Here the first distinctions between the sexes may be noted, and these are principally observable in the cocoons or puparia, rather than in the insect itself—at least to outward appearance. The male pupa is, in all cases—even in those where the female pupa is naked—enclosed in some kind of covering. In the *Diaspidinæ* the puparium is formed partly of fibrous secretion and partly of discarded skin; only, as the full-grown male emerges from it as a fly, and does not remain on the plant, there can be only one such skin—that of the larva; consequently it is easy to distinguish the male puparia from the shields of the adult females by the presence of only one discarded pellicle instead of two. In the *Lecanidinæ* and the *Coccidinæ* the male puparia are distinguishable usually by a narrower and more cylindrical form than those of the females, where these latter are covered; in the naked species the males are generally in white waxy or cottony cocoons.

Examination of the pupæ in these coverings will generally show more or less developed processes on the back and sides, which are so evidently the rudiments of the future wings that the presence of a male is not doubtful. In other respects the male pupæ are not always to be distinguished from the females.

3. *The full-grown male* has been described above. It is usually easy to procure specimens, provided the pupæ are obtained. If any of these, in their coverings, are put into pill-boxes with glass tops, or any place where light reaches them, they will generally produce the full-grown insect sometimes in a few days, sometimes after several weeks. The time of year for this seems very variable. Males emerge from the puparia apparently indifferently (in New Zealand) in summer or winter.

CHAPTER III.

PRODUCTS OF THE COCCIDIDÆ.

[Waxy or cottony matter: the "honeydew" and the black fungus—"smut" or "black blight"—growing upon it.]

THE Coccididæ, in some parts of the world, excrete various substances which are of commercial value, as, for example, shellac, "manna," candle-wax, &c. Cochineal is not in the same category, as it appears to be a colouring-matter pervading every cell of the tissues of the insect from which it is extracted —*Coccus cacti*. But there is no need to dwell here upon the ordinary excretions of the New Zealand insects, as they appear to be not sufficient either in quantity or quality for any practical service. The fibrous puparia of the *Diaspidinæ* appear to be quite useless. The tests of the *Lecano-diaspidæ*, such as *Ctenochiton perforatus*, *Inglisia ornata*, &c., although more or less waxy (but of very brittle material, often more like glass) are much too insignificant to repay any trouble taken to collect them. Of all the family, *Cælostoma zelandicum*, in its second stage, seems to produce the greatest amount of material, its large, hard, waxy tests being very thick and solid, and often clustered in hundreds on a root or a twig of Muhlenbeckia; but, supposing this substance (of which the true chemical nature* is not yet known) to be fit for some purpose, there does not seem to be any means of cultivating the insect to profit. *Dactylopius alpinus* produces in alcohol a rich red tint, and this not by way of excretion, but from the colouring matters of its tissues, as in the case of *Coccus cacti*; but here, again, the rarity of the insect and its out-of-the-way habitat would be a bar, even if nowadays it were worth while to cultivate a New Zealand cochineal. At present, therefore, there seems no reason to

---

* A small quantity was submitted to Mr. Skey, Colonial Museum Laboratory, for analysis, no more being available at the time. Mr. Skey considered it as a new substance, probably of the nature of a gum, not resinous; but further examination of larger quantities is necessary.

believe that the Coccididæ of this country are likely to furnish any products of a useful or commercial character.

There is, however, one substance produced by these insects which has an injurious effect upon the plants they grow on. This is a transparent glutinous fluid, apparently analogous to that exuding from Aphides, and which may receive the name of "honeydew," as in that family. In fact, this fluid would seem to be produced by most of the Rhynchota, for the Psyllidæ and Aleurodidæ also excrete it. The quantity issuing from Coccids seems to vary greatly. In some cases—e.g., *Lecanium hesperidum*, *Ctenochiton viridis* or *perforatus*, *Fiorinia asteliæ*—the insects appear to discharge "honeydew" freely; in others—e.g., *Mytilaspis pomorum*, *Rhizococcus fossor*—none, or scarcely any fluid, is excreted. But in no case does it appear that our Coccids* form honeydew to the same extent as the Aphides, which are stated to produce sometimes quantities that may be gathered from the leaves or the soil by the pound weight. It is not so much the amount exuding from each insect as the great number of insects on a plant which renders the Coccid honeydew obnoxious: each individual may excrete only a little, but when, as usually happens, there are many hundreds of individuals together, the result, for the reasons given below, becomes important to the tree.

There is every reason to believe that the honeydew of Coccididæ is of similar character to that of the Aphididæ, and, according to analyses by Boussingault, of Paris, and Gunning, of Amsterdam (Buckton, "Brit. Aphides," Vol. I., pp. 12, 13), the Aphidian honeydew contains a very large quantity of sugar, and, curiously enough, cane-sugar. Some observers, noticing in its composition also glucose and dextrine, have considered it as of vegetable rather than animal origin; but the weight of evidence appears to make it certainly the product of the Aphides. As the present work is intended rather as a manual for gardeners and tree-growers than as a purely scientific publication, there is no need to enter more fully into the subject here: it may therefore be simply stated that the honeydew of Coccididæ probably contains a large proportion of sugar in various forms.

The mode in which this substance is excreted by the insects differs somewhat from that of the Aphididæ. On the abdomen

---

* *Gossyparia mannipara*, an Arabian Coccid, is said to excrete so much that the Arabs "eat it with their bread like honey." Buckton, "Brit. Aphides," Vol. I., p. 42.

of Aphis are seen two erect more or less prominent tubes, called "cornicles" or "nectaries," and it is the function of these to excrete the honeydew.* No European entomologist has, it is believed, seen or described the organ of honeydew-excretion in the Coccididæ. Some observations by the author of this work in 1886 demonstrate its existence as a cylindrical tube exserted from the ano-genital orifice after the manner of a telescope, the furthest-extended tube being the most slender. This organ, extremely difficult of detection when not in use—except in the single genus *Cælostoma*—is at intervals pushed out to its full extent, and at its further extremity there appears a minute globule of yellowish, nearly transparent, glutinous fluid, which rapidly expands like a soap-bubble, and then, suddenly breaking, falls in spray on the leaf beneath. In the second stage of the female of *Cælostoma zælandicum* this organ may be detected more easily than in any other Coccid; but the act of protrusion of the organ and the formation of the drop of honeydew are apparently by no means frequent, and many long observations may be made without witnessing either.† (The organ and the honeydew-drop are shown in Plate xxii.)

For the purposes of this work further details as to the production of honeydew are not necessary. But as to its effect on plants it is requisite to be more particular, and the attention of tree-growers and gardeners is specially directed to the following points. It has been said above that when the bubble of honeydew has been expanded to its full size it breaks into spray. Now, as a general rule, Coccids are found almost exclusively on the *under* sides of leaves (when not on the bark). Some, as *Lecanium hesperidum* and a few others, may be seen on the upper side; but the general rule is as here stated. It follows that the spray of honeydew from the burst bubble falls, not on the leaf where the insect is, but on the *upper* sides of the leaves below it. These upper surfaces, being more exposed to light and air than the lower ones, are usually deserted not only by the Coccids but by other insects also, and so there is not much

---
* The fluid also emerges from the anal orifice; but, seemingly, no mention is made by any observer of any special honeydew-organ protruding from the anus of Aphis.

† Mr. Comstock ("Report on Insects," U.S. Dept. of Agric., 1881, p. 22) states that on gently rubbing a *Dactylopius* two small drops of fluid, which he considers to be honeydew, can be seen to emerge from orifices on the dorsal side of the sixth abdominal segment; but he mentions no special organ in the body The experiment has been tried on *Dactylopius* in this country without success.

chance that the honeydew spray so falling will be disturbed. It consequently rapidly accumulates and forms a coating on the leaf where it rests. From this result two things, or, rather, the same injurious effect on the plant is produced in two ways. First, the honeydew itself, being, as stated, of a glutinous nature, tends to stop up and choke the stomata (or, in plain English, the breathing-orifices of the leaves) and so retard the growth of the tree. Secondly, the honeydew, being of a saccharine nature, is especially attractive to fungoid growths, and these fungi, rapidly increasing, tend still more to choke the leaves and hamper the proper functions of the tree.

The second of these is the most important, for, apparently, the honeydew is scarcely deposited before it becomes the receptacle for fungus-spores, and these grow with great rapidity. As a general rule, in New Zealand, these fungi appear to be mostly of the same family—the Physomycetes, and they are of a black or very dark brown colour. From the fact above stated, that the honeydew falls from the insects upon leaves beneath them, the lower leaves of a plant are more covered with it than the upper ones: these black fungi consequently discolour chiefly the lower leaves and branches; often the uppermost branches are nearly free from them. But the effect produced on the tree is not only unsightly, from the sooty blackness, but also injurious, from the choking-up of the stomata both by the honeydew and the fungus. As for ornamental plants, whether under glass or in the open air, the black coating is quite sufficient to spoil them.

These fungi are of various species, and specimens are given here in Plate xxii.: on the leaves they form usually a hard, thin, black coat; while on the twigs and stem they are of a looser texture, forming masses of minute erect threads. They are not confined to New Zealand, and most writers on Coccididæ in Europe and elsewhere mention them, though only casually. They are, in fact, apparently, only the usual growths appearing on any decomposing substance, such as the honeydew is.

Gardeners and tree-growers ought to clearly understand that the appearance on their plants of this black sooty covering is almost always an indication of the presence of some Homopterous insects. In New Zealand, on account of the greater prevalence of Coccididæ, the insects will most probably be Coccids; but by no means necessarily so, for many Aphididæ,

Psyllidæ, and Aleurodidæ produce the same effects. This is by no means as well understood as it should be, either by gardeners themselves or by those who write on trees and planting. The fungus growth is usually imagined to be in itself a disease of the plant, and efforts are made for its treatment without regard to its real origin, the insects on the leaves or bark. Under the names of "smut," "black blight," &c., it is often referred to as a destructive pest; and remedies are suggested which can, of course, have no permanent effect unless they are equally efficacious against the insect producing the honeydew. It is probably from this cause that sulphur, which is an excellent remedy against fungus, has been so great a favourite with those who pretend to have discovered what are called "scaly-blight destroyers;" and gardeners, seeing, perhaps, these nostrums clean some of the fungus from their trees, are under the false impression that the "scale" is also cleared away. The truth is, that the real remedy against "black blight" is to kill the insects on whose excreta it flourishes, if that can be done. As to the modes of doing this see Chapter V.

It is not, of course, pretended here that fungi of different kinds, and even those specially referred to, will not grow independently of insects and honeydew; and trees are, undoubtedly, subject to fungoid diseases which are not to be traced to any animal action. Still, the rule holds good; and the first effort of a gardener on the appearance of black blight on his plants should be to discover the insects on its leaves or bark, and deal directly with them. Once they are destroyed the fungus growth will in a short time disappear.

CHAPTER IV.

CHECKS TO INCREASE OF COCCIDIDÆ, PARASITES, ETC.

The Coccididæ, like all Homoptera, produce great numbers of young; but their increase does not appear to be as rapid as that of some other families. The numbers of some Aphididæ or Aleurodidæ produced from a single female in the course of a single year have been calculated at hundreds of thousands, if not millions; and as many as eleven generations have been produced in little over half a year. Coccids, however, as a rule, do not propagate at this alarming rate. Many, if not the great majority of them, produce in this country but one generation in the year, e.g., *Mytilaspis pomorum*, *Cœlostoma zœlandicum*, &c. Others, such as *Icerya purchasi*, breed more often; and probably climate has a good deal to do with the frequency, for Mr. Comstock says that in the United States *Mytilaspis pomorum* breeds once a year in the North and twice in the South. In point of fact, it does not seem possible to lay down any rule on the subject. Unfortunately, *Icerya* is not only a frequent breeder, but also the most destructive insect of the family in New Zealand.

The number of young produced by each female seems also to vary. The author has counted from 30 to 80 eggs in the puparium of *Mytilaspis pomorum*; as many as 350 in the ovisac of *Icerya purchasi*, and about the same number of *Cœlostoma zœlandicum*; and a female of *Lecanium hesperidum* examined in spring contained 93 embryos. These figures do not denote any remarkable fertility; but, as in the majority of cases males are but seldom met with, sometimes even entirely unknown (e.g., *Lecanium hesperidum*), it follows that nearly every insect is capable of propagation, and the increase in numbers is therefore more rapid than might be anticipated otherwise. How the females in the species apparently destitute of males are enabled to produce young is perhaps one of the most mysterious things in Nature. The male of *Mytilaspis pomorum* has never been found in New Zealand or Europe, and doubtfully in America. *Lecanium hesperidum* has been known and studied for nearly two hundred years without any male, pupa or adult, being discovered. Yet both of these species go on increasing regularly and in great numbers, and show no signs of extinction.

In spite of this absence of males in some cases, and of the comparatively small numbers of eggs, Coccids would naturally increase at an exceedingly rapid rate if left undisturbed, on account of the great proportion of females. They are, moreover, protected, to a great extent—First, by the fact that birds do not, as a rule, care to eat them. The "blight-bird" or "white-eye," *Zosterops lateralis*, has been noticed in this country pecking about in holly-hedges infested by *Lecanium hesperidum;* but it is not absolutely certain whether it was eating the Coccids or the other more easily picked-off insects on the plant, such as *Psocus*, which is very commonly observed among Coccids. And other birds seem not to devour them at all. Secondly, the usual position of these insects, on the under side of the leaves, or in the crevices of bark, is a great shelter and protection for them against birds or ordinary accidents. Again, they are in many cases effectually covered by the waxy or fibrous shields, or by the masses of cotton with which they surround themselves. In countries like the South of France, California, or the greater part of New Zealand, the winters do not appear to be sufficiently severe to injure Coccids, and many of them breed as much in winter as they do in summer. It would therefore seem that everything combines to assist these insects in their career, and in their propagation. Nature, however, has provided a check which is to some extent effective, at least against several species, although, unhappily, against some of the most noxious—such as *Mytilaspis pomorum*, the Dactylopii (mealy bugs), *Icerya purchasi*, &c.—it is not energetic in this country; and this remedy is the attacks of other minute animals, whether by direct devouring of the Coccids or by parasitism.

Direct attacks from animal enemies are not frequent. Whether from some inherent distastefulness, or from the difficulty of getting at them, Coccids are scarcely subject to being directly devoured. There are a few exceptions. Under the puparia of *Mytilaspis pomorum* a minute white Acarus (mite) may often be found, and it is noticeable that where it abounds the eggs of the Coccid are often shrivelled or empty. It is not improbable that this Acarus may feed on the eggs. It appears to belong to the genus *Tyroglyphus*, a mite which is not usually carnivorous; but Mr. A. Michael, an authority on mites, seems to be uncertain whether *Tyroglyphus* may not make a meal of the Coccid.* No others amongst the Diaspidinae appear to be directly attacked, nor any of the Lecanidinae; but amongst a number of *Dactylopius glaucus* on a leaf there may often be seen a minute caterpillar, apparently covered with many tufts of hair. This is the larva of the common ladybird (*Coccinella*), a beetle which, in both the larval and perfect states, feeds on Aphides, Coccids, and other insects. The larva may sometimes be seen holding a *Dactylopius* in its powerful jaws and devouring it. Another larva, smooth and without hairs, performs the same functions—it is the young of a small dipterous fly, apparently one of the *Syrphidae*, insects also predatory; but this seems to be rare. In America, similar larvae are said to feed upon the "black scale" (*Lecanium oleae*). In this country, as far as is yet known, Lecanidae are not directly attacked by the ladybirds.

But this direct warfare by other animals is of small consequence. A more important check on the increase of many Coccids is afforded by the indirect action of minute hymenopterous insects, which make use of them as receptacles for their eggs. This plan is adopted by several Hymenoptera, of the families Chalcididae, Ichneumonidae, Proctotrupidae, &c. They do not devour their prey; they allow it to live that they may live on it. By means of their long ovipositor they pierce its body, and deposit in it an egg. As the victim grows the egg matures, changes into a larva, and still remains in the body of the Coccid; changes again into a pupa, and by this time the Coccid is at liberty to die, for the parasite has no further use for it except as a shelter; then, when the proper time arrives, the perfect fly emerges and departs. All Coccids are by no means equally subject to this system. In the United States, according to Mr. L. O. Howard,† parasites are found in nearly all genera—Diaspidinae, Lecanidinae, or Coccidinae. In this country, as far as observation shows at present, the Lecanidinae are the most

---

* Quart. Journ. Royal Micros. Soc., Feb., 1885.
† Report of the Entomologist, U.S. Dep. of Agric., 1880, pp. 350–371.

liable to attack, some of the Diaspidinæ next, and the Coccidinæ least of all. *Mytilaspis pomorum*, so hurtful to apple-trees, does not seem to be attacked. *Icerya purchasi*, the worst species of all, has not yet furnished a single parasitic fly. A few specimens of *Dactylopius glaucus* contain parasites; a good many of *Fiorinia asteliæ* and *F. stricta*; while *Ctenochiton perforatus* and *C. viridis* appear to be the favourite victims, as sometimes scarcely any specimens on a plant can be examined which do not contain either a larva or a pupa of a parasite. It is to be observed that in no case is more than one parasite to be found in a Coccid.

Unfortunately, although this provision of Nature must have a very considerable effect in preventing the increase of Coccididæ, it is subject to two drawbacks. In the first place, as just observed, some of the most injurious pests appear to be unaffected by it. Probably, up to the present time the worst plant-enemies* in New Zealand have been *Mytilaspis pomorum*, *Aspidiotus coccineus*, *Aspidiotus camelliæ*, *Lecanium hesperidum*, *Lecanium oleæ*, *Lecanium hemisphæricum*, *Pulvinaria camellicola*, and *Icerya purchasi*. Here reference is made not so much to insects which render plants unsightly as to those which seriously injure its growth: many others, such as *Fiorinia asteliæ* or *Ctenochiton viridis* are ugly enough, but have not been destructive. Of the injurious species above named none, apparently, are troubled in this country by parasitic insects up to the present time, at least to any appreciable extent.

A sketch of Ctenochiton enclosing a parasitic pupa, and of the perfect fly, will be found in Plate xxiii. In a work like this the generic and specific characters of these parasites need not be given: they do not seem to differ much from hymenopterous and dipterous insects of other countries.

Another mode by which the too rapid increase of Coccids is checked is by the attacks of vegetable parasites—fungoid growths which permeate the whole body of the insect, and soon kill it. As far as experience in New Zealand extends as yet the genera *Ctenochiton*, *Lecanium*, and probably *Eriochiton* are the only ones so attacked. On certain plants in the forests, notably Hedycarya and Coprosma, circular spots may be commonly found on the under side of the leaves: some dark-

---

* Speaking of Coccids only; *Kermaphis pini* is equally, if not more, destructive.

brown, somewhat convex, some bright yellow and often quite globular. In spring, examination of a young larva of *Ctenochiton viridis*—a species very common on the above plants—will frequently show, either within the insect, or on its waxy test, or between the test and the insect, minute specks, which under a high power of the microscope, prove to resemble the filaments composing the brown or the yellow spots just mentioned. On turning over one of the brown fungi, or on pulling it to pieces, the dead body of a young *Ctenochiton* or *Lecanium* larva will always be found in the middle. Apparently this brown fungus does not attack any but young larvæ; but the bright yellow fungus will be found filling the bodies also of the females in the second stage, and the globular portion of the fungus will stand out above them. These fungi are not of the same genus as *Empusa*, the fungus which so frequently kills the house-fly; but they seem to act in much the same way within the insect.

Probably a good many of the *Lecano-diaspidæ* are preyed on and destroyed by these fungoid parasites, of which figures are given in Plate xxiii.

CHAPTER V.

REMEDIES AGAINST COCCIDIDÆ.

Many people are under the impression that scale-insects out-of-doors are not of much consequence. They are aware that in greenhouses and hothouses these insects are a trouble to gardeners, and that they probably injure flowering or fruit-bearing plants in such situations. But they imagine that in the open air, and on large well-grown trees, Coccids do no very great harm; or, if the trees are for a time injured, that recovery and health will come before long, and the pest will disappear. This is not the place in which to controvert this or any other opinion. A work professedly dealing with facts should be as free as possible from controversial discussion. Whatever, therefore, may be the grounds of the opinion just stated, or the reasons for rejecting it, it will be sufficient here to say that there seems to be nothing to lead to the belief that New Zealand is likely to be different from other countries in this respect. To institute a comparison, it would be manifestly absurd to include such countries as England, or Germany, or, on the other hand, India, or Central America, or North Australia—Firstly, because in the greater part, or at least in the northern parts, of Europe the winters are much more severe than in New Zealand, and almost certainly the great cold is injurious to such insects as Coccids. Secondly, because in tropical countries it seems that the too great heat is equally obnoxious to them; and, with the exception of a few species, tropical Coccids are comparatively harmless. But it is to the warmer temperate or the subtropical regions that we must look for comparison—regions where there is neither too scorching a summer nor too ice-bound a winter. And, for this purpose, we have only to take such lands as California, Florida, the South of France or Northern Italy, the Cape of Good Hope, the southern regions of Australia, &c. The experience of these is, that some species of Coccids do injure, in every way, whether as regards ornamental or commercial value, a number of trees and plants on which the people of the country depend largely for subsistence or profit. In the South of France the olive industry has been in some years greatly cut down. In Florida, California, Australia, the Cape of Good Hope, oranges and apples have been so damaged that the value of an orchard or a grove has been reduced sometimes by 80 per cent. It may be said, moreover, that even in tropical countries the attacks of scale-insects are often most damaging: in Mauritius the sugar-cane and in Ceylon the coffee plantations have suffered from their ravages. The experience of American fruit-growers is certainly not to be despised, and the fact that both in California and Florida the people strain every nerve to get rid of the insect pests on open-air trees would seem to be distinctly against the notion that these little enemies can be neglected with impunity.

Nor, indeed, can it be said that in New Zealand itself the attacks of scale-insects out-of-doors are harmless. Apple orchards throughout the country bear evidence to the contrary: lemon-groves can be seen about Auckland where, instead of the thousands of fruit formerly grown, a few stunted lemons are all that the withered trees afford; and nobody can glance round the plantations at Nelson or Napier without recognizing the devastating powers of a scale-insect (*Icerya*).

The opinion that Coccids are not likely to do much harm in the open air is therefore scarcely tenable, and it will be of use to inquire what remedies can be provided against them.

There is a point, however, to be noted at the outset, and it is, that in reality there is not, as far as is yet known, any *certain* remedy against scale-insects. Not that ingenuity and experiment have not succeeded in inventing plans and substances quite efficient in killing the insects when applied to them. It is easy enough to kill an insect when you can get at it, in most cases; but the problem in this instance is not only to kill individual insects, but to do more. What is wanted is to get rid of whole communities of them, and, at the same time, to prevent their eggs from hatching and a new brood coming forth. Many of those who profess to know all about destroying "scale"— especially if they belong to that class which prides itself upon being "practical men"—being generally quite ignorant of the habits and life-history of the insects, are satisfied when they have tried some rule-of-thumb plan which seems to kill most of the adult insects, not dreaming that they have left the eggs unharmed and ready to send forth a fresh swarm at hatching-time. There is another obstacle which often prevents success in eradicating "scale." This is the difficulty of making sure of the effects of any remedy. A plan which has answered well in one place will fail in another, and this, not only as regards different countries, but even in the same district for neighbouring gardens, or even for neighbouring trees in the same garden. Tree-growers must be prepared to find the very same remedy which has cleared their neighbours' trees fail for their own; and in this country the author has seen, in one and the same orchard, some trees quite cleared, while on others, treated in exactly a similar manner, the "scale" was scarcely injured.

It is from want of knowledge of this and the like points that persons who have tried various remedies recommended to them have complained of failure, and condemned both the remedy and their adviser, whilst really neither was in fault.

Nothing need be said here of carelessness or unskilfulness in applying a remedy, beyond the following instance: A person whose apple-trees were being very much damaged by *Mytilaspis pomorum* was advised to apply, by way of painting the trunks and branches, a mixture of kerosene and some other ingredient. In two or three months he found violent fault with his adviser, for he said every tree was dead or dying. On inquiry it was found that, from over-zeal or want of knowledge, he had applied the mixture as if painting a house, had used it much too strong, and, to make assurance doubly sure, had given his trees two good coats of the oil.

An intelligent appreciation of the life-history and habits of scale-insects is necessary to enable any one to select and apply, with a probability of success, a remedy, and the details given in Chapter I. of this work will be found useful for this purpose. It will be apparent from them that, without regarding the generic or specific characters of these insects, we may lay down a few general principles to start with, thus:—

I. *Whatever damage is done is effected by the sucking of the juices of the plant through the rostrum of the insect.* It follows from this that applications of any fluid to the tree externally, with the object of poisoning the insects in their feeding, would be useless, as their food is drawn from beneath the surface.

II. Neglecting entomological distinctions, we may divide the Coccididae, roughly, into—
(*a*.) Insects attacking deciduous plants;
(*b*.) Insects attacking evergreen plants;
  or, again,
(*c*.) Insects living usually on the bark;
(*d*.) Insects living usually on the leaves;
(*e*.) Insects living on both bark and leaves;
  or, lastly,
(*f*.) Insects covered with hard shields or "scales;"
(*g*.) Insects covered with cotton;
(*h*.) Insects naked.

It will be clear that a different method will be required for destroying these different classes; but any one insect will belong to more than one class. Thus *Mytilaspis pomorum*, the apple-scale, belongs to (*a*), (*c*), and (*f*), and indeed may be placed also in (*b*), as such plants as hawthorns, which it attacks, are as bad as evergreens in the difficulty of reaching the insect on them; or, *Lecanium hesperidum* is in (*a*), (*d*), and (*h*); *Lecanium oleae* in (*a*), (*b*), (*e*), and (*h*).

As far as regards the injurious species of Coccids it may usually be taken for granted that those infesting deciduous plants (class *a*) live chiefly on the bark (class *c*), and are either naked (class *h*) or covered with a hard scale (class *f*). If naked they are chiefly *Lecanium*; if covered, either *Mytilaspis*, *Aspidiotus*, or *Diaspis*.

Icerya is exceptional, being omnivorous, feeding equally on bark or leaves, deciduous or evergreen plants; it belongs to every class except (*f*). Every method of destruction has therefore to be resorted to against it.

The treatment of a deciduous plant infested by Coccids is simple as to its method. For two reasons the dead winter-time must be chosen for it—first, because, the leaves being off, the whole plant can be easily got at; secondly, because the eggs of the insect have not yet been hatched, and the whole brood can be destroyed at once. The first operation should be the pruning of the tree, so as to reduce the labour required to a minimum. A brushing over all the trunk and branches with a good hard stiff brush and one of the liquid remedies given below is then often successful. Brushing with a *dry* brush is adopted by some persons; but, although this doubtless clears away a good many insects and scales, and may do the tree itself some good by cleaning off fungus-growths and incrustations, yet it necessarily fails to destroy all the eggs, and in consequence the work is only half done. *Any one who wishes to extirpate Coccids must make certain that he has destroyed the eggs*—a fact which is quite ignored by numbers of those who glibly talk of their own success, and blame the advice of others. The object being, therefore, twofold, the operation should be performed with a hard, stiff brush dipped in one of the fluids recommended below; and care should be taken that there is no part of the trunk or branches escaping untouched. In fact, what should be aimed at is a kind of painting of the tree, but with a thin coating of the fluid, so as to close the pores as little as may be; while at the same time the brush clears away as many as possible of the "scales" and their enclosed broods of insects and eggs.

Bearing in mind what has been said just now of the want of *certainty* in any remedy whatsoever, the tree-grower who follows these directions will most likely find his work successful and his deciduous plants cleaned of "scale" on the bark.

A second method may be adopted—namely, the painting-over of the trunk and branches, without attempting to forcibly detach any "scales" with the brush. This, *properly performed and with proper fluids*, is likely to be just as efficacious as the other, for the fluid should "run in" under the scales, surround the eggs, and prevent them from hatching. It gives less trouble than the hard brushing, and is equally destructive to the Coccids. It has, however, of course, not the same cleaning effect upon fungoid growths or incrustations impeding the free "respiration" of the plant.

For deciduous trees, then, such as apple- or pear-trees in an orchard, the simple remedy is severe pruning at the dead of winter, and the coating of the trees with a destructive fluid, laid on with a brush *on every part*, preferably with a hard brush vigorously used, but leaving a thin coat of the fluid on the bark.

It must be thoroughly understood that, a week or two after the first application, the "scales" left on the tree should be examined, and, if the eggs are not killed, a second coating of the fluid should be applied.

The treatment of evergreen plants, or of plants which are attacked both on the bark and leaves, is really the same as the above as regards its object, but it necessarily differs in its method. Here, again, it is desired not only to kill the insects themselves, but also to devitalize the eggs; but in this case the work is much harder, for the eggs are especially difficult to reach. Still, there is this advantage: that in dealing with evergreens the season of the year need not be specially studied, and, in default of touching the eggs, one may get at the young larvæ. The remedy is again a fluid, but it must be applied in the form of spray. Coccids are sometimes found on the upper surfaces of leaves, but as a rule they affect the lower sides. This, of course, renders it much more difficult to get at them; and the ordinary rose of a garden syringe would not, as a general thing, distribute the fluid in a sufficiently-fine form. The finer the spray and the more it is forced into every corner and nook of the plant the better. Various force-pumps and spray-throwers have been invented for this purpose in the United States; but probably tree-growers in this country need only procure the finest possible rose for their syringes, and use them in the ordinary way. The fluid should be thrown as thoroughly as possible on all parts of the plant, every care being taken to direct it most fully against the under sides of the leaves.

The work, then, to be done is in itself simple enough. A destructive fluid must be selected and applied according to the character of the insect and its position on the tree. For covered or naked insects on the bark, apply it with a hard, stiff brush; for covered or naked insects on the leaves, apply it in the form of the *finest spray* thoroughly forced as much as possible into every nook and cranny, and especially against the under side of the leaves.

The question, "What is the best fluid to use?" is more complicated. Many answers have been given to it: many fluids have been strongly recommended by different people. It must be well remembered that, as stated above, a sure and sovereign remedy has yet to be discovered, and failure may attend even the best suggested at present. Bearing this in mind, tree-growers will find in the following list the result of the experience of a number of observers, which may serve as a useful guide. It does not profess to be more than a summary, compiled from the researches of entomologists such as Mr. Comstock, Professor Riley, Mr. Hubbard; from suggestions by gardeners and others, embodied in various parliamentary and private documents; and from actual observation and experiment in this country: but it is believed that the information here given may be accepted and relied on.

Some of the substances here given are manifestly unsuitable for general use on account of their expense, at any rate in the open air. Yet it is well to include them, as they are all suggested in some work or other, or in the replies of gardeners and fruit-growers to parliamentary inquiries; and the objections to them ought also to be known :—

1. *Alcohol.* Will certainly kill any individual insect; but "sprayed over scale-insects produced no apparent effect" (Comstock).
2. *Ammonia.* Whether used pure (diluted) or in urine, damages the plants much more than it does the insects (Hubbard; Comstock).
3. *Ashes.* Powdered, or mixed with lime, salt, soot, &c. Of no value whatever (Hubbard; Personal experiment).
4. *Carbolic acid.* Of no avail, either as spray or brushed on, unless used in such strength as to seriously damage the tree (Hubbard; Riley; Comstock).
5. *Castor-oil.* Has been found efficacious in cleaning hawthorn-trees at the Agricultural College, Lincoln (T. Kirk). It was mixed with soot for some unexplained reason. The time of the year when it was applied is not stated; but the author's experiments seem to show that castor-oil *does not effectually kill the eggs.* Still, it is doubtless a valuable remedy if applied repeatedly, so as to kill larvæ and adults, supposing it to be sufficiently cheap.
6. *Cole's Insect-exterminator.* Apparently a mixture of about 2 parts of "green soap" with 100 parts of strong alcohol. It is "effectual as an insecticide, and harmless to growing plants;" but "the cost is too great, except on a small scale, as in conservatories" (Comstock).
7. *Gasoline.* Seems to have been used in California on pear-trees: result, doubtful (New Zealand Parliamentary Papers: Codlin Moth Committee Report, 1885, page 8).
8. *Gishurst compound.* Very favourably spoken of in many quarters. In New South Wales it has been found efficacious on orange-trees against *Aspidiotus coc-*

cineus (Alderton); in Nelson it is said to be used beneficially against *Icerya purchasi. It does not, however, kill the eggs with certainty* (Personal experiment). Applied warm, and properly diluted, it may be recommended as a good remedy; but applications of it should be repeated.

9. *Kerosene.* Seemingly the most valuable of all remedies, when properly applied. " Almost the only substance which will certainly kill the eggs without at the same time destroying the plant" (Hubbard).

But the application of this remedy must be carefully performed. Some trees may endure it without injury, even undiluted or unmixed; but this is scarcely to be expected, and the oil should therefore be applied in some mixed form. Also, it is important to remember *that a hot sun increases the injurious effect of kerosene*: consequently winter, or cloudy weather, should be chosen for its employment.

(*a.*) *Pure kerosene.* As just stated, it is probably not advisable to use this. Still, "a young shoot of orange, not more than fourteen days old, was uninjured by an application of pure kerosene which thoroughly wet every leaf;" (Comstock); and *Lecanium hesperidum* on ivy, similarly treated, was destroyed, without injury to the plant (ibid.).

(*b.*) *Kerosene and milk.* An excellent mixture, if milk can be obtained cheap (Riley; Hubbard; Comstock). It must be applied in the form of an "emulsion," sprayed over the tree or brushed on the bark. Hubbard gives the following directions for use: Heat the milk nearly to boiling-point and mix with double the quantity of kerosene; churn violently from ten minutes to half an hour, according to temperature, until a creamy thick fluid is obtained; dilute this with nine or ten times the quantity of water. The mixture is of course purely a *mechanical* one, as far at least as the water is concerned, and it must be kept constantly stirred, to prevent the substances from separating from the water. For evergreen trees impel the mixture on leaves and branches *in the finest possible spray.* Sour milk is as useful as fresh.

The object of the milk is not only to lessen the injurious qualities of the kerosene, but also to induce it to mix more freely with the water; but it is the oil alone which destroys the insects *and their eggs*.

(*c*.) *Kerosene and soap*. When milk is not obtainable, or too dear, nothing is so excellent as this mixture. Soap itself (see below) is a useful insecticide, and in combination with kerosene includes the good qualities of both substances. The cheapest possible qualities of soap will do. The mixture, which is, even more than the last, purely *mechanical*, must be made first of all an "emulsion." The American experiments result in the following recipe and method of using :—

Formula :

Common soap	½lb.
Kerosene	2 galls.
Soft water	1 gall.

Dissolve the soap in the water heated to boiling, then add the kerosene, and churn the mixture until a creamy fluid results which thickens on cooling. Dilute with nine or ten times the quantity of water: the quantities given above will make about thirty gallons of liquid. Whale-oil soap, soft-soap, or any other kind will do. As with the milk emulsion, apply in the form of the finest spray for evergreens (Riley ; Hubbard ; Personal experiment).

(*d*.) *Kerosene and oil*. Castor-oil, linseed-oil, whale-oil, may be used. A mixture of this kind, in the proportion of 1 part kerosene to 3 or 4 of oil, has been found very efficacious for apple- and other fruit-trees attacked by the common apple-scale *(Myt. pomorum)*. But, as observed above, the mixture must not be laid on too thick. Thinly brushed all over trunk and branches, at dead of winter, it has been found quite successful in destroying both insects *and eggs*, without injury to the trees (Personal experiment). It would probably not answer for evergreens, on account of expense.

On the whole, it may be said that, as far as certainty can be attained in the matter, there is no substance better for destroying Coccids *and their eggs* than kerosene in the form of milk or soap emulsion, diluted with water for evergreens or for trees with insects on the leaves as well as on the bark. Probably, for deciduous fruit-trees the kerosene-and-oil mixture is the best.

The great point in favour of this substance is that *it destroys the eggs*; this, few if any of the others will accomplish.

10. *Lime.* Of no avail whatever.
11. *Lye.* Concentrated lye is very frequently recommended. In the New Zealand Parliamentary Papers (Codlin Moth Committee Report, 1885, page 7) several statements will be found apparently most favourable to it; yet in places we find admissions that " it cannot reach all the eggs." In America generally, it has not been found satisfactory : "inferior to kerosene in killing-power, and far more injurious to trees when used in solutions strong enough to be effective as insecticides."

It is quite possible that the action of lye on the fungus accompanying the scale-insects (see Chap. III.) may have led "practical" gardeners to imagine that it cleaned their trees of scale. Comstock says, " I saw most excellent results from the following mixture : 1lb. concentrated lye, one pint gasoline or benzine, half pint oil, five gallons water." Probably the good results here were due, not to the lye, but to the gasoline and oil.

12. *Pyrethrum.* Useless against Coccids (Comstock).
13. *Salt.* Useless (Comstock).
14. *Soap.* Undoubtedly a valuable remedy, and perhaps, in some cases, as efficient as kerosene; *but it does not destroy the eggs.* A solution of ¾lb. of soap to a gallon of water, *applied hot*, was entirely successful in California : three months after its application no living scale-insect could be found (Comstock). The time of the year is not stated. In another case the solution was applied cold : "four days after no living insect could be found ;" but, again, the time of the year is not stated, and no mention is made of the eggs. Still, a strong solution of soap may be said

to be one of the best remedies against the larvæ and adult insects—proportions from ¼lb. to ½lb. soap to one gallon of water (Comstock; Hubbard; Personal experiment).

15. *Soda, caustic.* Strongly recommended by many persons. It injures the tree, *and does not kill the eggs*—two things which are decidedly against its use. Gardeners may have been led to employ it from finding that in some instances it clears away the black fungus-growths (Chap. III.), and imagining this to be a clearance also of the scales.

16. *Soda, silicate.* Kills some insects, *but no eggs*, and injures the tree (Hubbard).

17. *Sulphate of iron.* "A common ingredient in patent remedies;" most injurious to vegetation. It does not affect scale-insects (Hubbard).

18. *Sulphur.* Another substance, the object of a kind of superstitious veneration amongst gardeners. It is excellent against fungoid growth, but of little value against scale-insects. Here, again, the clearing of the black fungus has probably been taken to mean also the destruction of the insects (Hubbard; Comstock; Personal experiment). Comstock says that in America people often bore holes in their trees and stuff them with sulphur, under the notion that the substance will be taken up by the sap, and poison the insects: quite a futile idea.

19. *Sulphur and lime.* A dangerous compound, and useless unless applied in such strength as to kill the tree. Its fumes are poisonous, and it may seriously injure the face and hands (Hubbard).

20. *Sulphur and snuff.* Equal parts mixed and dusted over *Lecanium hesperidum* on a wet day were quite successful (Comstock). But the mixture would be too expensive except for conservatory plants, and doubtless the snuff alone would be quite as efficacious.

21. *Sulphuric acid.* "Killed nearly all the scale-insects, and very nearly killed the tree" (Hubbard). No mention is made of its action on the eggs.

22. *Soot.* Useless (Hubbard; Comstock; Personal experiment).

23. *Tobacco.* A good remedy against larvæ and adults; *doubtful against the eggs.* Fumigation has no effect on scale-insects, except sometimes on Dactylopidæ, or "mealy bugs" (Hubbard; Comstock; Personal experiment). The tobacco should be applied in a pretty strong solution; but the expense in this country would be probably too great for general use.

24. *Whale-oil and whale-oil soap* have been already alluded to under the head "Kerosene." They are both useful ingredients in mixture with that substance, if procurable cheaply.

From the foregoing list it will be gathered that, if experiment, combined with knowledge of the habits and life-history of scale-insects, can be relied on, there is no substance better adapted for their destruction than kerosene, mixed with oil, or milk, or soap solution, and *carefully applied.* It has been already observed that the killing of the eggs is absolutely necessary for thorough clearing-away of the insects; and, to quote again the words of Mr. Hubbard, kerosene is "almost the only substance which will with certainty kill the eggs without at the same time destroying the plant."

But precautions must not be neglected. Persons who recklessly use any remedy, or who apply it too thickly or in too strong proportions, must expect their trees to suffer. Nor must the weather and the time of the year be overlooked. *Winter is the best season for all remedies;* and, preferably, cool and cloudy days. Again, if substances soluble in water, such as potash or soda lye, soap solutions, &c., be employed, it must be expected that a day's rain will wash a good deal of them off, and greatly reduce their efficacy. These are things which many people forget; they fancy that because somebody has cleared his trees with, say, castor-oil in winter they can do the same thing in full heat of summer; or, because a lye solution has done well in the dry climate of California, that it will be equally good in the rains of New Zealand. Still more is it a fallacy to imagine that rule-of-thumb methods, not founded upon any knowledge of the nature, habits, and life-history of the insects, are likely to be really efficacious.

Little need be said here of a remedy which has had, to some extent, the authority of Professor Riley, and which is recommended by Mr. Howard (Report U.S. Dep. of Agric. 1880–81,

p. 351): viz., the transportation or acclimatization of parasites on scale-insects. Doubtless the thing could be done, as experiments in America have shown. But there are plenty of parasitic insects in New Zealand already, and, although they seem to have hitherto confined their work to the native and mostly to the innoxious Coccids, they may at any time begin to attack the others, and it is only a question of time when they will act usefully as efficient checks (see Chap. IV.).

There is one Coccid of which it must be said that, whilst kerosene mixtures will undoubtedly destroy it, by far the best remedy of all is to destroy and burn at once the infested trees. *Icerya purchasi* is so voracious and universal a feeder, so repulsive in its aspect, and so destructive in its effects that the most drastic remedy is the best. Any one, therefore, having a tree, especially an ornamental or a fruit tree, attacked by *Icerya purchasi*, is strongly recommended to make no delay, but to cut down and burn every stick of the tree as soon as possible.

It was observed at the beginning of this chapter that some people hold the opinion that the damage done by scale-insects is not of importance. The foregoing remarks upon remedies are not directed to those who hold this view, which is contradicted by the experience not only of other countries but of New Zealand itself.

Authorities referred to in this Chapter.

U.S. Department of Agriculture—
    Reports by Professor Riley, Professor Comstock, Mr. Hubbard, Mr. L. O. Howard.
N.Z. Parliamentary Papers, 1885—
    Report of the Select Joint Committee of both Houses on the Codlin-moth, and "various blights to which fruits are subject."
Personal experiment by the author and friends.
Replies of farmers, gardeners, and tree-growers to inquiries, official or private.

CHAPTER VI.

CATALOGUE OF INSECTS AND DIAGNOSIS OF SPECIES.

Since this work has been in type, the author has received a letter from the State Inspector of Fruit Pests for California, in which the writer states that the insect *Icerya Purchasi* has there, especially in the southern part of the State, gained such hold on the orange-groves, in spite of the most strenuous efforts, that the people find it impossible to keep it down. Orange- and lemon-growers (and indeed other tree-growers) in New Zealand, especially in the North Island, should take note of this fact, and beware of ever letting this omnivorous and most destructive insect obtain any footing on their trees. *A speedy burning of every infected tree is the best remedy in this case.*

present in larva and pupa, always absent in adult; tarsi monomerous; feet ending in a single claw; abdomen terminating in a spike which forms the sheath of the penis; eyes present in adult; ocelli often large, sometimes exceeding three in number.

The above characters sufficiently distinguish this family from the rest of the Homoptera. Probably the first marks for identification of a specimen might be the monomerous tarsus and the single claw. The latter is always to be made out, at least in the earlier stages of the female and in the adult male.

GROUPS.

Larvæ active, naked; adult females and pupæ stationary, covered with separate shields or puparia, composed partly of secretion, partly of the earlier discarded pellicles; females apodous after larval stage; abdomen of females not exhibiting a median cleft or dorsal lobes DIASPIDINÆ.

p. 351) : viz., the transportation or acclimatization of parasites on scale-insects. Doubtless the thing could be done, as experiments in America have shown. But there are plenty of parasitic insects in New Zealand already, and, although they seem to have hitherto confined their work to the native and mostly to the innoxious Coccids, they may at any time begin to attack the others, and it is only a question of time when they will act usefully as efficient checks (see Chap. IV.).

There is one Coccid of which it must be said that, whilst kerosene mixtures will undoubtedly destroy it, by far the best remedy of all is to destroy and burn at once the infested trees.

U.S. Department of Agriculture—
    Reports by Professor Riley, Professor Comstock, Mr. Hubbard, Mr. L. O. Howard.
N.Z. Parliamentary Papers, 1885—
    Report of the Select Joint Committee of both Houses on the Codlin-moth, and " various blights to which fruits are subject."
Personal experiment by the author and friends.
Replies of farmers, gardeners, and tree-growers to inquiries, official or private.

CHAPTER VI.

CATALOGUE OF INSECTS AND DIAGNOSIS OF SPECIES.

Family.—COCCIDIDÆ.

MALE and female larvæ similar, apterous, naked or covered, active.

Females in all stages apterous; metamorphosis semi-complete; naked or covered; active or stationary; rostrum usually present in all stages, sometimes absent in adult; feet sometimes absent after larval stage; tarsi where present monomerous; feet, where present, ending in a single claw; eyes sometimes absent.

Male pupæ apterous; naked or covered. Adult males with two wings and two halteres; metamorphosis complete; rostrum present in larva and pupa, always absent in adult; tarsi monomerous; feet ending in a single claw; abdomen terminating in a spike which forms the sheath of the penis; eyes present in adult; ocelli often large, sometimes exceeding three in number.

The above characters sufficiently distinguish this family from the rest of the Homoptera. Probably the first marks for identification of a specimen might be the monomerous tarsus and the single claw. The latter is always to be made out, at least in the earlier stages of the female and in the adult male.

GROUPS.

Larvæ active, naked; adult females and pupæ stationary, covered with separate shields or puparia, composed partly of secretion, partly of the earlier discarded pellicles; females apodous after larval stage; abdomen of females not exhibiting a median cleft or dorsal lobes   DIASPIDINÆ.

Larvæ active, naked; adult females and pupæ active or stationary, naked or covered with secretion; adults sometimes apodous; abdomen of females exhibiting a median cleft and two dorsal lobes   LECANIDINÆ.

Larvæ active, naked, exhibiting at the abdominal extremity two protruding anal tubercles. Adult females exhibiting abdominal cleft and dorsal lobes; naked or covered with secretion   HEMICOCCIDINÆ.

Females in all stages exhibiting anal tubercles; no abdominal cleft or dorsal lobes; naked or covered with secretion   COCCIDINÆ.

Group I.—DIASPIDINÆ.

Female insects covering themselves with separate shields or puparia composed partly of fibrous secretion, partly of the discarded pellicles; females apodous after first stage; no abdominal cleft or lobes; spinnerets usually arranged in groups on the posterior segment of female.

GENERA.

Female puparium circular, pellicles usually in the centre; male puparium slightly elongated, not carinated, pellicle at one end. Four or five groups of spinnerets, or groups absent — ASPIDIOTUS.

Female puparium more or less circular, pellicles near the centre; male puparium elongated, carinated, pellicle at one end. Five groups of spinnerets — DIASPIS.

Female puparium elongated, pellicles at one end; male puparium nearly similar but smaller and narrower, not carinated, pellicle at one end. Five groups of spinnerets — MYTILASPIS.

Female puparium elongated, pellicles at one end; male puparium much narrower and smaller, carinated, pellicle at one end. Five groups of spinnerets — CHIONASPIS.

Female puparium elongated, pellicles at one end; male puparium narrower, pellicle at one end. More than five groups of spinnerets. Abdomen of female not fringed — POLIASPIS.

Female puparium elongated, first pellicle at one end, second pellicle almost filling the puparium; male puparium smaller and narrower, sometimes carinated, pellicle at one end — FIORINIA.

Genera not yet represented in New Zealand.

Female puparium circular or elongated; male puparium elongated, not carinated — PARLATORIA.

Female puparium elongated; male puparium similar but smaller. Abdomen of female fringed LEUCASPIS.

Female puparium double, the scales superimposed, first pellicle in the centre of the upper scale; male puparium elongated, not carinated AONIDIA.

Female puparium completely enclosing the insect; male puparium elongated, not carinated TARGIONIA.

Genus: ASPIDIOTUS, Bouché.

Female puparium varying in colour; circular in outline, usually flat, sometimes rather convex; pellicles usually in the centre.

Male puparium rather longer than that of the female, the pellicle at one end; not carinated above.

Groups of spinnerets usually four, sometimes five, and in one American species (*A. sabalis*, Comstock) six; or, in a few cases, altogether wanting.

Adult females usually peg-top shaped.

1. ASPIDIOTUS ATHEROSPERMÆ, Maskell.
    N.Z. Trans., Vol. XI., 1878, p. 198.
    (Plate IV., Fig. 1.)

Female puparium circular, flat, brown; diameter, about $\frac{1}{20}$ in. The pellicles in the centre form sometimes a sort of boss or protuberance, of lighter colour than the rest.

Male puparium oval, flat, brownish, about $\frac{1}{36}$ in. in length.

Adult female light-yellow in colour, of the usual peg-top shape of the genus; length, about $\frac{1}{35}$ in.; corrugated—the last abdominal section, being very small, is much overlapped by the rest. Four groups of spinnerets, upper pair with fifteen orifices; lower pair, nine or ten. Abdomen terminating in several lobes, of which the four median are the largest; between the lobes scaly serrated hairs.

Adult male unknown.

Habitat—On *Atherosperma Novæ Zealandiæ*, Wellington; Hawke's Bay.

2. ASPIDIOTUS BUDLÆIÆ, Signoret.
    Ann. de la Soc. Entom. de France, 1868, p. 115.
    N.Z. Trans., Vol. XI., 1878, p. 198.

Female puparium circular, flat, dirty-white, about $\frac{1}{15}$in. in diameter.

Male puparium oval, dirty-white, about $\frac{1}{30}$in. in length.

Adult female light-yellow, peg-top shaped; abdomen ending in two somewhat prominent lobes, with scaly hairs and spines. Four groups of spinnerets: upper groups with five or six orifices; lower groups, three or four.

Adult male yellow, slightly brown on the thorax; antennæ of ten joints, all hairy.

Habitat—On silver-wattle, Nelson. The insect is European, and found there on *Buddlæia salicina*.

Closely allied to *A. nerii*, but differing in the lobes and spinnerets of the abdomen.

3. ASPIDIOTUS CAMELLIÆ, Boisduval.

Kermes camelliæ, Boisduval, Ent. Hort., p. 334.
Aspidiotus camelliæ, Signoret, loc. cit., 1869, p. 117.
N.Z. Trans., Vol. XI., 1878, p. 200; Vol. XVII., 1884, p. 21.

(Plate IV., Fig. 2.)

Female puparium nearly circular, convex, greyish or brownish in colour, about $\frac{1}{15}$in. in diameter; pellicles often at one side.

Male puparium rather smaller, oval.

Adult female of normal shape, but somewhat elongated. Abdomen ending in two lobes, with a few scattered scaly hairs. No groups of spinnerets.

Habitat—On camellias, Christchurch; on euonymus, weeping willow, &c., Wellington.

Very common in gardens about Wellington: sometimes does much damage to euonymus shrubs and hedges.

4. ASPIDIOTUS CARPODETI, Maskell.

N.Z. Trans., Vol. XVII., 1884, p. 21.

Female puparium usually light-brown, but varying a little with the colour of the bark of the tree; convex; circular; the pellicles in the centre; some specimens are slightly elongated. Average diameter, $\frac{1}{16}$in.

Male puparium narrow, with parallel sides; not carinated; dirty-white or brownish colour; length, about $\frac{1}{16}$in.

Adult female of the normal peg-top shape, the abdomen not so much overlapped as usual. Abdomen ending in two median somewhat prominent lobes, with two others much smaller not in close proximity; edge of the body jagged, with curvilinear incisions, amongst which and between the lobes are a number of serrated pointed hairs, as in *A. nerii*. Four groups of spinnerets: lower pair with four to six orifices; upper, with six to ten. These groups seem surrounded by a narrow line as if enclosed in a chamber: the same appearance is presented (according to a figure of Mr. Comstock's) in *A. nerii*. There are many single spinnerets.

Adult male of normal form, with antennæ of ten joints, of which the seventh, eighth, and ninth are the longest. The haltere has a somewhat long peduncle. The abdominal spike is rather long, and springs from a large tubercle.

Habitat—On *Carpodetus serratus* and *Vitex littoralis* (puriri), Wellington. The puparia are so like in colour to the bark that it is difficult to detect them.

This insect is evidently closely allied to *A. nerii*, but differs in the abdominal lobes of the female and in the antennæ of the male; its male puparium is also much longer, and that of the female more convex, than in that species.

5. ASPIDIOTUS COCCINEUS, Gennadius.*

Aspidiotus aurantii, Maskell.
N.Z. Trans., Vol. XI., 1878, p. 199.
Aspidiotus citri, Comstock; Canadian Entom., Vol. XIII., p. 8.

(Plate IV., Fig. 3.)

Female puparium really dirty-white, but seeming yellowish-brown, from the colour of the insect beneath; sometimes dark-brown; circular, flat; diameter, about $\frac{1}{11}$ in.

Male puparium much smaller, rather oval.

Adult female yellow, becoming brown at last; peg-top shaped, but the abdominal segment is comparatively so small and is so much overlapped by the others that the insect looks almost globular; length, about $\frac{1}{25}$ in. Abdomen ending in six lobes (of which the two median are the largest), and several scaly hairs. No groups of spinnerets.

---

* The author has not been able to find the original description of Gennadius, which appears to have been contained in a report to the Minister for Agriculture in Greece. Dr. Signoret states that there is a reference to it in "Risso, Histoire Naturelle des Oranges," Vol. I., p. 220.

Adult male very small, brown or yellow in colour. The antennæ have ten joints : the two first joints are very small, round, and smooth; the third, fourth, fifth, and sixth equal in length; the seventh, eighth, and ninth half as long; the tenth somewhat shorter still, and pointed. All the last eight joints show numerous hairs. The thorax is short and thick, the thoracic band occupying more than one-half the width; the abdomen short, the double spike of some length. The wings are oval, about as long as the body. The legs are hairy, femora thick, tibiæ longer, thicker at the end next the tarsus than at the other end; tarsi broad at the top, tapering gradually down to the usual single claw. The hairs on the femora are much fewer than those on the tibiæ and tarsi.

Habitat—On oranges and lemons in shops, very abundant, often several hundreds on a single fruit ; on orange- and lemon-trees, Governor's Bay, Canterbury ; and Auckland.

This insect is European, and has been introduced here from Australia. It is exceedingly destructive to orange and lemon groves in America and Australia. Mr. Comstock (Report of the Entomologist, U.S. Dep. of Agric., 1881, p. 295) records an instance where a grove of thirty-three acres, which in 1872 produced a rental of £1,800, could fetch in 1878 only £120, on account of the ravages of this insect.

Orange- and lemon-growers in the north of New Zealand should beware of this pest. It is scarcely likely that it should be harmless here when it is so destructive elsewhere.

The remedies most likely to be efficacious have been mentioned in the introductory chapters of this work.

6. ASPIDIOTUS DYSOXYLI, Maskell.

N.Z. Trans., Vol. XI., 1878, p. 198.

Female puparium circular, somewhat convex, brown in colour ; diameter, about 1/5 in.

Male puparium smaller, oval, brown.

Adult female bright-yellow, corrugated, the corrugations overlapping the abdominal region, which is comparatively small. There are four groups of spinnerets, the upper pair with ten openings, the lower with nine, many scattered oval and oblong spinnerets. The abdomen ends in six lobes, of which only the two median are conspicuous ; between the lobes fine serrated hairs. The abdomen is very velvety.

Adult male unknown.

Habitat—On *Dysoxylon spectabile*, Wellington.

Allied to *A. atherospermæ*, but differing in the abdominal lobes and spinnerets.

7. ASPIDIOTUS EPIDENDRI, Bouché.
    Chermes epidendri, Boisduval ; Ent. Hort., 1867, p. 339.
    Aspidiotus epidendri, Signoret, loc. cit., 1869, p. 121.
    N.Z. Trans., Vol. XI., 1878, p. 197.

Female puparium circular, flat, dirty-white or brownish; diameter, about $\frac{1}{12}$in.

Male puparium elongated, the sides parallel.

Adult female greenish yellow, peg-top shaped. Abdomen ending in several lobes, of which only the two median are conspicuous; between the lobes several serrated scaly hairs, and some spines. Four groups of spinnerets: upper groups, eight to ten orifices; lower groups, six to eight: many single spinnerets.

Adult male somewhat long, yellowish in colour; antennæ of ten joints; feet having somewhat thick femora, the tibiæ and tarsi slender; all the joints hairy. The abdominal spike, or sheath of the penis, is rather long, and the tubercle at its base is large.

Habitat—On palms and orchids in hothouses, *passim*; on wattle, rarely, Christchurch.

This is a European insect, affecting hothouse plants, and scarcely likely to do damage out-of-doors. It is closely allied to *A. nerii*.

8. ASPIDIOTUS NERII, Bouché; Schadl. Gart. Ins., 1833, 52.
    Diaspis Bouchei, Targioni-Tozzetti ; "Studie sulle Cocciniglie," 1867.
    Aspidiotus Bouchei, Targioni; Catal., 1868, 45, 1.
    N.Z. Trans., Vol. XIV., 1881, p. 217.
    (Plate IV., Fig. 4.)

Female puparium circular, flat, white or greyish; diameter, about $\frac{1}{12}$in.

Male puparium oval, white; about $\frac{1}{25}$in. in length.

Adult female yellow, peg-top shaped, but almost globular. Abdomen ending in six lobes, of which the two median are the largest. Between and a little beyond the lobes are a number of scaly serrated hairs, some of which exhibit serrated extremities;

also some scaly but smooth hairs. There are also a few spines. Four groups of spinnerets, which are surrounded (according to Mr. Comstock, Entom. Report, U.S. Dep. of Agric., 1880, Plate XV., Fig. 1) by a fine line, as if enclosed in a sac. Many single spinnerets.

In the larva the abdomen ends in four lobes, of which the two median are somewhat prominent.

Adult male yellow or brownish; antennae of ten joints, each with several hairs; feet having the femora somewhat thick, the tibiae and tarsi flat and slender, the former a little expanded at the extremity.

Habitat—On *Coprosma lucida* and *Corynocarpus lævigata* (Karaka), Wellington.

A species introduced from Europe, where its favourite habitat is *Nerium oleander*; but it is found on many other plants, and is, according to Dr. Signoret, "the commonest of all the species of this genus." It has not yet spread widely in New Zealand.

9. ASPIDIOTUS SOPHORÆ, Maskell.

N.Z. Trans., Vol. XVI., 1883, p. 121.

Female puparium nearly circular, flat, bluish-grey; diameter, about $\frac{1}{24}$in.

Male puparium oval; length, about $\frac{1}{36}$in.

Adult female of the usual peg-top shape, greenish-yellow in colour; abdomen ending in two conspicuous median lobes, with a number of scaly serrated hairs, as in *A. nerii*. Five groups of spinnerets: uppermost group with four orifices; the remainder, seven or eight. Some specimens show only four groups.

Adult male unknown.

Habitat—On *Sophora tetraptera* (Kowhai), Port Hills, Canterbury.

Only a few species of Aspidiotus are reported with five groups of spinnerets. The present insect differs from all of them in the scaly serrated hairs of the abdomen; none of the others has more than a few spines.

Genus: DIASPIS, Costa.

Female puparium more or less, but never quite, circular; sometimes flat, but more usually convex; pellicles more or less marginal.

Male puparium elongated, the pellicle at one end; a longitudinal carina, or keel, appears in the middle.

Groups of spinnerets, five.

Mr. Comstock (Entom. Rep., Cornell Univ., 1883, p. 85) remarks that, when the pellicles of the female of this genus are marginal, it might be difficult to distinguish it from Chionaspis, as the male puparia are alike in both. As regards the species observed hitherto in New Zealand this difficulty has not occurred.

10. DIASPIS BOISDUVALII, Signoret; Ann. de la Soc. Entom. de France, 1868, p. 433.

N.Z. Trans., Vol. XI., 1878, p. 200; Vol. XVII., 1884, p. 23.

(Plate IV., Fig. 5.)

Female puparium oval, nearly circular, flattish; colour, yellowish-grey; diameter, about $\frac{1}{12}$ in.

Male puparium elongated, white, with a strong median keel, and with the edges raised so as to appear like two other keels; length, about $\frac{1}{20}$ in. The male puparia frequently occur massed in great numbers, and covered with white hairs and fluff.

Adult female rather elongated, oval, or somewhat pear-shaped; the body corrugated, the cephalic portion smooth. At each side, on a level with the rostrum, or a little above it, is a protruding lobe, which is characteristic. Colour, light-yellow. Abdomen ending in two lobes, not prominent, and with a depression between them; beyond the lobes are many serrations, with small lobelike projections and spiny hairs. Five groups of spinnerets: uppermost group with five to eight orifices[*]; the two upper side groups with twenty to twenty-five; lower side groups, fifteen to twenty. A few scattered single spinnerets.

Adult male very small, about $\frac{1}{40}$ in. in length; colour, reddish-yellow; antennæ of ten joints, all with hairs except the two first; femora and tibiæ slender, tarsi thick at the base, and tapering to the claw; digitules, fine hairs. The first and second pair of legs appear somewhat widely separated, owing to the length of the coxæ.

Habitat—On several hothouse plants, Christchurch and Wellington; and on wattle in gardens, Wellington.

A European insect. The curious projections at the side,

---

[*] Mr. Comstock (Entom. Rep., Cornell Univ., 1883, p. 87) gives eight to fifteen orifices for the uppermost group.

near the head, of the female, and the arrangement of the male puparia above mentioned, sufficiently distinguish this species.

11. DIASPIS ROSÆ, Sandberg.

Aspidiotus rosæ, Sandberg; Abhand., priv. Boh., No. 6, p. 317.

Diaspis rosæ, Signoret, loc. cit., 1869, p. 141.

N.Z. Trans., Vol. XI., 1878, p. 201.

(Plate IV., Fig. 6.)

Female puparium nearly circular, white, often aggregated in masses; diameter, about $\frac{1}{12}$in. Pellicles, marginal.

Male puparium white, elongated, carinated; length, about $\frac{1}{20}$in.

Adult female deep-red in colour, elongated, the body deeply segmented. Cephalic region very large, smooth. On each segment of the body several spiny hairs. Abdomen ending in two conspicuous lobes with a depression between them, and some spiny hairs. Five groups of spinnerets, but the lateral groups are almost continuous; uppermost group with about twenty orifices; in the lateral groups, fifty to sixty orifices. No single spinnerets.

Adult male orange-red in colour; antennæ ten-jointed, with several hairs on all but the first two joints; feet slender, hairy; digitales, fine hairs. The spike is somewhat long.

Habitat — On rose-trees, Governor's Bay, Canterbury—Napier.

A European species, stated by Mr. Comstock to attack, in America, blackberries and raspberries, besides the rose.

The deep-red colour and abnormally-large cephalic segment of this insect distinguish it from all others.

12. DIASPIS SANTALI, Maskell.

N.Z. Trans., Vol. XVI., 1883, p. 122.

(Plate IV., Fig. 7.)

Female puparium yellowish-grey in colour, sometimes with a greenish tinge; outline oval; very convex; pellicles at one end, black, inconspicuous; length of puparium, about $\frac{1}{15}$in.

Male puparium white, elongated, carinated; pellicle, black; length, about $\frac{1}{25}$in.

Adult female orange-red in colour, peg-top shaped; the abdominal segment very small as compared with the rest of the body, and the two next segments overlap it. Abdomen ending in two conspicuous, prominent, median lobes, and at each side of them two semi-circular depressions: several branched and serrated hairs in the region of these lobes. There are no groups of spinnerets. There is no wide depression of the edge between the median lobes.

Adult male unknown.

Habitat — On *Santalum cunninghamii* (Maire), Te Aute, Hawke's Bay; and on pear, plum, and other fruit-trees at Whangarei, having probably spread from native plants.

The carinated male puparium distinguishes this species from Aspidiotus. The absence of spinnerets is curious.

Genus: MYTILASPIS, Targioni-Tozzetti.

Female and male puparia similar, or nearly similar, in shape, but the male puparium is smaller. Puparia elongated, generally more or less mussel-shaped or pyriform, usually convex, more or less curved; pellicles at one end. Male puparia not carinated. Five groups of spinnerets in the female, but the groups are sometimes continuous.

13. MYTILASPIS CORDYLINIDIS, Maskell.

N.Z. Trans., Vol. XI., 1878, p. 195.

(Plate V., Fig. 1.)

Female puparium pure white, elongated, very narrow; usually straight, sometimes curved; pellicles yellow, at one end; length, about $\frac{1}{6}$in.; breadth, about $\frac{1}{30}$in.

Male puparium similar to that of the female, but much smaller; length, about $\frac{1}{25}$in.

Adult female pale yellow in colour, elongated, distinctly segmented. Rudimentary antennæ on the cephalic segment. A few fine hairs at the edges of the segments. Abdomen ending in two lobes with a small median depression; several serrated scaly hairs, and a few spines. Five groups of spinnerets: uppermost groups, seven or eight orifices; upper lateral group, fourteen to twenty; lower lateral group, twenty to twenty-five. A great number of single spinnerets.

Adult male doubtful; very minute and difficult to detect. Antennæ apparently short and tibiæ large.

Habitat—On *Cordyline australis* and *C. indivisa*, *Phormium*, *Gahnia*, *Astelia*, *Eucalyptus*, &c., throughout the islands; but the chief habitat seems to be *C. australis* (the common cabbage-tree), on which it is often very abundant.

This species may at first sight be mistaken for *Fiorinia stricta*, described below, which also infests Cordyline and Phormium; but, on inspection, it will be seen that the puparium of the Mytilaspis is much whiter, and the pellicles yellow, those of *F. stricta* being black. An examination of the second pellicles of the two species will, of course, at once distinguish them.

14. MYTILASPIS DRIMYDIS, Maskell.

N.Z. Trans., Vol. XI., 1878, p. 196.

(Plate V., Fig. 3.)

Female puparium elongated, often straight, sometimes curved; colour, dirty-white or brown; pellicles at one end; length, about $\frac{1}{12}$in.

Male puparium similar, but smaller.

Adult female dull-red in colour, elongated, not very distinctly segmented. Abdomen ending in a number of small lobes, of which the four median are the largest; several fine hairs between the lobes; no groups of spinnerets, but a very great number of single ones, which are scattered on the segments as far up as the rostrum. Many of these protrude as short thick tubes with serrated or fringed extremities. On the cephalic segment are a few spiny hairs and two rudimentary antennæ.

Adult male red in colour; antennæ of ten joints; tarsi somewhat large. Both antennæ and feet have numerous hairs. Digitules, fine.

Habitat—On *Drimys colorata*, Water of Leith, Dunedin, from which it has spread to other native plants.

15. MYTILASPIS EPIPHYTIDIS, Maskell.

N.Z. Trans., Vol. XVII., 1884, p. 21.

(Plate V., Fig. 2.)

Female puparium flat, pyriform, brown in colour, thin; length, about $\frac{1}{11}$in.

Male puparium narrower than that of the female, and a good deal darker, being sometimes almost black; length, about $\frac{1}{20}$ in.; not carinated.

Adult female dark-grey in colour, elongated, segmented. Abdomen ending in two median lobes; along the edge several deepish curvilinear incisions, between which are some strong spines. Five groups of spinnerets: lower pair with fourteen to sixteen orifices; upper pair, twelve to sixteen; uppermost group, four to six.

Adult male unknown.

Habitat—On *Astelia cunninghamii*, Wellington.

16. MYTILASPIS LEPTOSPERMI, Maskell.

N.Z. Trans., Vol. XIV., 1881, p. 215.

(Plate V., Fig. 4.)

Female puparium flat, elongated, irregularly pyriform, light-brown in colour; length, about $\frac{1}{12}$ in. The secretion forming the puparium is mixed with bark-cells of the tree, arranged longitudinally.

Male puparium narrower than that of the female, and darker in colour.

Adult female greyish-green, elongated, segmented; abdomen ending in six lobes, of which the two median are conspicuous and somewhat large and floriated, the rest very small. Five groups of spinnerets: the upper group with about fifteen openings; the others with from twenty-five to thirty-five. Single spinnerets none, or very few.

Adult male unknown.

Habitat—On *Leptospermum scoparium* (manuka), Wellington; Canterbury; Auckland. The puparia are often numerous on the loose scaly bark of the tree.

17. MYTILASPIS METROSIDERI, Maskell.

N.Z. Trans., Vol. XII., 1879, p. 293.

Female puparium white, pyriform. Female in all stages dark-coloured; in last stage nearly black. General outline resembling *M. drimydis*, but the abdomen is much sharper and more pointed, with a finely-serrated edge, ending in three minute, pointed lobes, joined by a scaly process. Spinnerets in an almost continuous arch, which may be resolved into five groups; seventy or eighty openings; several single spinnerets. The rudimentary antennæ can be made out.

The young female has an elongated oval outline, little corrugated. The feet, digitules, antennæ, &c., resemble those of *M. pomorum*. The abdomen is like that of the adult, without the groups of spinnerets.

Male unknown, but puparium smaller and rather darker in colour than that of the female.

Habitat—On *Metrosideros robusta* (rata), Wellington, and probably elsewhere. It is not common.

18. MYTILASPIS PHYMATODIDIS, Maskell.

N.Z. Trans., Vol. XII., 1879, p. 292.

Female puparium flattish, pyriform, dirty-white or brownish; length, about $\frac{1}{12}$in.

Male puparium similar, brown.

Adult female greyish, elongated, segmented. Rudimentary antennæ visible. Abdomen ending in two lobes with a median depression: several scaly and serrated processes, and some spiny hairs. Five groups of spinnerets: uppermost group, six to nine orifices; upper side groups, ten to fourteen; lower pair, fifteen to twenty: several single spinnerets.

Male unknown.

Habitat—On *Phymatodes billardieri*, Wellington; Auckland.

In outward appearance the female resembles *M. pomorum*, but the puparium is quite different, and the abdominal characters also differ.

19. MYTILASPIS POMORUM, Bouché.

Aspidiotus pomorum, Bouché; Ent. Zeit. Stett., 1851, XII., No. 1.

Aspidiotus conchiformis, auctorum: nec Gmelin, Syst. Nat., 2,221.

Aspidiotus pyrus-malus, Kennicott; 1854, Acad. Science of Cleveland.

Aspidiotus juglandis, Fitch; Ann. Rep., N.Y. State Ag. Soc., 1856; nec Signoret, loc. cit., 1870, p. 95.

Aspidiotus falciformis, Bärensprung; Journ. d'Alton et Burm., 1849.

Mytilaspis pomicorticis, Riley; Fifth Rep. State Entom., Missouri, p. 95.
Mytilaspis pomorum (Bouché), Signoret; loc. cit., 1870, p. 98.
N.Z. Trans., Vol. XI., 1878, p. 192.
The common apple-scale.

(Plate V., Fig. 5.)

Female puparium usually brown,* sometimes white; elongated, mussel-shaped, convex, slightly curved, sometimes straight; length, about $\frac{1}{10}$ in.

Male puparium not known in New Zealand. In America it is stated† to be small, "straight or nearly so, and with the posterior part joined to the remainder of the scale by a thin portion which serves as a hinge."

Adult female greyish, yellowish, or white; elongated, segmented. Rudimentary antennæ present. At the edge of each segment two or three strong spines. Abdomen ending in two large lobes, with two others much smaller on each side; the median lobes are trifoliated. Between and beyond the lobes some spines. Five groups of spinnerets; numbers of orifices variable (see below); a few single spinnerets.

Male unknown in New Zealand and Europe, doubtful in America. Colour stated by Riley (Fifth Missouri Report, p. 95) as " translucent corneous-grey."

Habitat in New Zealand—On apple, pear, plum, peach, apricot, lilac, ash, thorn, sycamore, cotoneaster, and other plants, *passim.*

An introduced European species, known in America and elsewhere as the "oyster-shell bark-louse of the apple." It is the commonest, apparently, of the Diaspidinæ; and does great damage in orchards.

This species has been referred to by many writers under the specific name "conchiformis;" some authors include it under the genus Aspidiotus, others under Coccus, and one—Réaumur —under Chermes. In the Quarterly Journal of the Royal Microscopical Society, February, 1885, Mr. A. Michael refers to it as *Coccus (Mytilaspis) pomicorticis.*

The groups of spinnerets have been stated above to be

---

* Dr. Signoret says, "brun noirâtre." Mr. Comstock calls it "ash-grey." In reality the colour varies somewhat with that of the bark of the tree.
† Comstock; Rep. of Entom., U.S. Dept. of Agric., 1880, p. 325.

"variable." The following table shows the numbers observed in specimens from different trees in New Zealand:—

	Uppermost Groups.	Upper Side Groups.	Lower Side Groups.
Apple	17	17	14
Plum	20	17	17
Lilac	17	19	16
Ash	10	12	9
Cotoneaster	7	15	10

A very minute white Acarid (mite) has been observed frequently under the puparia of this species, among the eggs. The eggs, in most cases so observed, were shrivelled and dead. Mr. A. Michael, in the paper above mentioned ("Notes on Tyroglyphidæ") refers to an Acarus found in America in 1873, also in puparia of *M. pomorum*, by Mr. Riley, and expresses doubts whether or not it fed upon the insect; yet he says, "A Tyroglyphus not ordinarily predatory might regard a Coccus as suitable for gastronomic purposes."

20. MYTILASPIS PYRIFORMIS, Maskell.

N.Z. Trans., Vol. XI., 1878, p. 194; Vol. XIV., 1881, p. 215; Vol. XVII., 1884, p. 22.

(Plate V., Fig. 6.)

Female puparium light-brown, elongated, pyriform, flat; length, averaging $\frac{1}{10}$in. (sometimes reaching $\frac{1}{8}$in.); breadth, averaging $\frac{1}{25}$in. (reaching $\frac{1}{20}$in.); texture, thin. The second pellicle is comparatively large.

Male puparium smaller and narrower, brown, not carinated.

Adult female yellowish-brown or greyish; elongated, segmented; on the segments are a few spiny hairs. Abdomen ending in several lobes, of which the two median are much the largest. Spinnerets in a continuous arch, containing sixty to seventy orifices. Many single spinnerets. Several scaly hairs between the lobes.

Adult male orange-coloured, about $\frac{1}{30}$in. long. Antennæ 10-jointed. Digitules, long fine hairs. Sheath of the penis long.

Habitat—On *Dysoxylon spectabile* and *Atherosperma Novæ Zælandiæ*, Wellington; on *Coprosma*, Riccarton Bush, Canterbury.

In the female puparium and in the length of the abdominal spike of the male this species resembles *M. buxi*, Bouché (Signoret, loc. cit., 1868, p. 93), but differs in all other respects.

Genus: CHIONASPIS, Signoret.

Female puparium usually white, elongated; pellicles at one end; generally flat.

Male puparium white, elongated, carinated; pellicle at one end.

Groups of spinnerets, five (in one American species, six); rarely wanting.

21. CHIONASPIS CITRI, Comstock; 2nd Rep., Dep. of Entom., Cornell Univ., 1883.
    Chionaspis euonymi, Comstock (in part); Ag. Rep., 1880, p. 313.
    N.Z. Trans., Vol. XVII., p. 1884, p. 23.

(Plate VI., Fig. 1.)

Female puparium dirty blackish-brown, with a grey margin; elongated. "There is a central ridge from which the sides slope like the roof of a house" (Comstock).

Male puparium white, narrow, carinated.

Adult female yellowish-white, elongated, segmented. Abdomen ending in six lobes, of which the two median are the largest: these two are divergent. Along the edge some spines. No groups of spinnerets: a few single ones.

Adult male unknown.

Habitat — On oranges sold in the shops, imported from Sydney.

This insect, apparently an importation from America, was not observed prior to 1884, and occurs as yet only sparingly, mingled with *A. coccineus*, from which it is easily distinguished by its elongated puparium.

22. CHIONASPIS DUBIA, Maskell.
    N.Z. Trans., Vol. XIV., 1881, p. 216.

(Plate VI., Fig. 2.)

Female puparium white, flat, elongated, pyriform, very thin; the pellicles rather small; length, about $\tfrac{1}{12}$ in.

Male puparium white, elongated, rather oval; very slightly carinated above; on the under-side are two longitudinal keels.

Adult female yellow, elongated, segmented; the abdominal segments somewhat deep. Abdomen ending with a median depression; terminal lobes inconspicuous (absent?). Five groups of spinnerets: uppermost group, six to ten orifices; the rest, ten to fifteen.

Adult male reddish in colour. Antennæ hairy, 10-jointed, the first two joints very short. Feet normal, with four long, fine digitules. At the base of the abdominal spike is a somewhat large tubercle. Haltere of normal form, but the terminal seta is very long, four times as long as the thick basal portion, and has no terminal knob. Thoracic band conspicuous. The thorax is somewhat long, so that there is a considerable distance between the first and second pairs of legs.

Habitat—On *Coprosma*, *Rubus*, *Asplenium*, *Pellæa*, Riccarton Bush and North Kowai River, Canterbury; Auckland.

The female puparium resembles that of *C. aspidistræ* (Signoret) and *C. populi* (Bärensprung), but the abdominal segment of the female differs from both.

23. CHIONASPIS DYSOXYLI, Maskell.

N.Z. Trans., Vol. XVII., 1884, p. 22.

(Plate VI., Fig. 3.)

Female puparium thin, flattish, pyriform, white in colour, with a faint pink tinge when the egg-mass beneath shows through it; length, about $\frac{1}{12}$ in. The second pellicle is comparatively large.

Male puparium white, narrow, carinated; length, about $\frac{1}{36}$ in.

The insect affects principally the leaves of the plant, and the puparia are usually clustered thickly along the midrib.

Adult female yellowish-red, elongated, segmented; but not very deeply. Abdomen ending in a broken curve, with many curvilinear incisions. There are fourteen lobes, of which the two median are the largest; separated from them by a spine on each side are two others rather smaller; then another spine and a short open space; and then three smaller lobes and another spine; another space, and then a single small lobe, followed by a spine. Five groups of spinnerets: lower pair with twelve to fourteen orifices; upper pair with seven to ten; uppermost group, four to six. A few spiny hairs are on the edge of the abdomen.

Adult male unknown.

Habitat—On *Dysoxylon spectabile* (Kohe-kohe), Wellington; Hawke's Bay; Auckland.

The large white puparia of this insect do much to spoil the appearance of Dysoxylon, one of the most showy-leafed plants in New Zealand.

24. CHIONASPIS MINOR, Maskell.

N.Z. Trans., Vol. XVII., 1884, p. 33.

(Plate VI., Fig. 4.)

Female puparium white, small, not more than $\frac{1}{15}$in. in length, usually less; it is narrower and less pyriform than is usual in the genus, and is often bent in the middle; pellicles yellow.

Male puparium white, narrow, elongated, carinated, about $\frac{1}{30}$in. in length.

Adult female elongated; segmented, but not deeply; colour, dark-brown. Abdomen ending in six small lobes, of which the two median—the largest—are closely contiguous. Between them and the next pair is a spine; then beyond the second pair another spine, a space, and a third pair of very small lobes; after a long space there is another spine. Five groups of spinnerets: uppermost group with twelve to fourteen orifices; upper pair, fourteen to seventeen; lower pair, eighteen to twenty-four: many single spinnerets.

Adult male not known.

Habitat—On *Parsonsia*, Hawke's Bay; on *Rhipogonum scandens* (supplejack), Wellington; Canterbury; Otago.

The small puparia and the contiguous abdominal lobes of the female distinguish this species.

Genus: POLIASPIS, Maskell; N.Z. Trans., Vol. XII., 1879, p. 293.

Female puparia elongated; pellicles at one end. Male puparia narrower, elongated, pellicle at one end. Female with more than five groups of spinnerets; abdomen without fringe.

In the kindred genus, *Leucaspis*, Targioni-Tozzetti (Signoret, loc. cit., 1868, p. 101), the abdomen has a continuous fringe of long spines, and the groups of spinnerets vary in number from five to eight.

25. POLIASPIS MEDIA, Maskell.
N.Z. Trans., Vol. XII., 1879, p. 293.
(Plate VI., Fig. 5.)

Female puparium white, elongated, pyriform, slightly convex; length, about $\frac{1}{18}$ in.

Male puparium elongated, narrow, white, doubtfully carinated.

Adult female elongated, segmented; greenish-white; length, about $\frac{1}{24}$ in. Rudimentary antennæ visible. Abdomen ending with a median depression, and inconspicuous lobes; several scattered spiny hairs. Eight groups of spinnerets: four, containing each from twenty to thirty orifices, are placed in opposite pairs, the fifth, with four to six orifices, being between the upper pair; above these, three other groups form an arch, the two outer ones having eight to ten openings, the middle one three to five. Many single spinnerets.

Adult male of a bright scarlet or deep-orange colour. The antennæ, covered with longish hairs, have ten joints, the first two very short and thick; the next five long, equal, and cylindrical; the eighth and ninth somewhat shorter; the tenth fusiform, and as long as the seventh. The legs are rather long; the femur thick, the tibia more slender, broadening towards the tarsus, which is about one-third as long as the tibia, and narrows sharply down to the claw. Both tarsus and tibia are hairy. The digitules are fine hairs.

Habitat—On *Veronica*, sp., and *Leucopogon Fraseri*, North Kowai River, Canterbury; on *Cyathodes acerosa*, Wellington; on ferns, Napier.

Genus: FIORINIA, Targioni-Tozzetti. UHLERIA, Comstock;
2nd Entom. Rep., Cornell Univ., 1883, p. 110.

Female puparium elongated; first pellicle small, at one end; second pellicle very large, entirely covering the insect, and almost extending to the edges of the puparium.

Male puparium elongated; smaller and narrower than that of the female; sometimes carinated; pellicle at one end.

Mr. Comstock proposes the name "Uhleria" for this genus, because Professor Targioni, establishing his genus for the species to which he originally gave the name of *Diaspis fiorinæ*, changed at the same time the specific name to "pellucida."

This, Mr. Comstock says, necessitates now an entirely new generic name.

Targioni's nomenclature has been followed here, as likely to lead to less confusion.

26. FIORINIA ASTELIÆ, Maskell.
   Diaspis gigas, Maskell.
   N.Z. Trans., Vol. XI., 1878, p. 201; Vol. XIV., 1881, p. 217; Vol. XVII., 1884, p. 24.
   Uhleria gigas, Comstock; 2nd Entom. Rep., Cornell Univ., 1883, p. 111.
   (Plate VI., Fig. 6.)

Female puparium elongated, flat, roughly pyriform or ovate, thin; the secretion is yellowish-brown or dirty-white, but is scarcely noticeable, on account of the second pellicle; length variable, from $\frac{1}{12}$in. to $\frac{1}{8}$in.; breadth, about $\frac{1}{16}$in. First pellicle small, at one end. Second pellicle, very large, almost filling the puparium, roughly pyriform; abdominal region segmented; cephalic region large, oval; abdominal segments tapering, exhibiting at the extremity either minute serrations, floriated lobes, or tusk-like lobes, or a smooth curve; the first abdominal segment sometimes produced into roundly-triangular lobes.

Male puparium flattish, elongated; length, from $\frac{1}{16}$in. to $\frac{1}{8}$in.; white, thin; roughly pyriform, but narrower than that of the female; central portion slightly convex, seeming on the underside to have two keels; not carinated above.

Adult female yellow or brown; segmented; at first elongated, the cephalic region comparatively large, but during gestation shrinking up until the insect assumes the form of Aspidiotus. Abdomen ending in a minutely-serrated edge, with several small simple lobes, between which are longish spiny hairs. Spinnerets in an almost continuous arch, containing seventy to a hundred orifices; several single spinnerets.

Adult male yellow, slender. Antennæ, ten-jointed, as long as the body; each joint except the two first long and hairy; the last joint fusiform. Feet, long and slender; digitules, fine hairs. Abdominal spike, slender, not very long, springing from a small tubercular base.

This is a variable insect in size, colour, edge of abdomen, and spinnerets. On the bark of *Pittosporum eugenioide* a variety has the extremity of the second pellicle richly floriated, other features remaining as above. It has not been thought advisable to erect all these varieties into different species.

The male pupa, in its earlier state, is elongated, segmented, and may be mistaken for a female of *Mytilaspis drimydis*: but differs in its greyish-yellow colour, and also in the form of the puparium.

Habitat—On *Atherosperma Novæ Zealandiæ*; *Astelia cunninghamii*; *Coprosma*, sp. var.; *Pittosporum eugenioide*, &c.; Wellington; Canterbury; Hawke's Bay; Otago; Nelson; Auckland.

The puparia of this insect are frequently covered by a species of torulaceous (?) fungus which spreads over the leaf they are on in a thin, brown sheet.

27. FIORINIA GROSSULARIÆ, Maskell.

N.Z. Trans., Vol. XVI., 1883, p. 123.

Female puparium irregularly oval, being formed chiefly by the second pellicle, with a narrow edge of fibrous secretion; length, about $\frac{1}{20}$in.

Adult female elongated; segmented; the cephalic end slightly prolonged into a compressed cylinder. Segments bearing at the edge three sharp spines. The edge of the abdominal segment is much broken by serrations, and ends in two broadish median lobes, with two smaller lobes on each side. Several sharp, long spines are set in pairs along the serrated edge. Five groups of spinnerets, the three upper forming a continuous arch. Colour of insect, dark-grey.

Adult male and puparium unknown.

Habitat—On gooseberries, Amberley, Canterbury.

A doubtful species.

28. FIORINIA MINIMA, Maskell.

N.Z. Trans., Vol. XVI., 1883, p. 122.

Female puparium flat, elongated, oval; length, about $\frac{1}{16}$in. First pellicle, comparatively large; the second almost fills the puparium.

Male puparium rather longer than that of the female, but much narrower; carinated.

Adult female elongated; segmented; colour, pink. The abdominal segment somewhat long, the edge broken by a number of deepish curvilinear serrations, and ending in two inconspicuous median lobes, with three others, much smaller, on each side. From the serrations spring some hairs. There are five groups of spinnerets, but the three upper ones, almost or quite conjoined, form a nearly-continuous arch, containing forty to fifty orifices; the two lower groups have fifteen to twenty. There are several single spinnerets. The adult insect, before gestation, nearly fills the space covered by the second pellicle; after gestation it shrinks up into very small compass at the cephalic end of the puparium.

Adult male unknown.

Habitat—On *Brachyglottis repanda*; *Panax arboreum*, Port Hills, Canterbury.

Differs from the European species, *F. pellucida* (Targioni), in its extremely-minute size, in the serrations of the abdomen, and the number of its hairs. Also in *F. pellucida* the young female larva has two tubercles between the antennae, which are not seen in *F. minima*.

29. FIORINIA STRICTA, Maskell.
 N.Z. Trans., Vol. XVI., 1883, p. 124; Vol. XVII., 1884, p. 24.

(Plate VI., Fig. 7.)

Female puparium elongated, narrow, with sides almost straight and parallel; length, about $\frac{1}{14}$in.; breadth, about $\frac{1}{50}$in. Colour of secretion, white, but seeming black, as the second pellicle shows through it. First pellicle, black, small, at one end; the cephalic portion prolonged in a slightly-cylindrical form. Second pellicle, very long, filling the puparium; black; entire for most of its length, but at the abdominal extremity cut across by transverse divisions forming narrow radiating segments; extreme edge semicircular, sharply serrated. Texture, horny, hard, and strong.

Male puparium elongated, narrow, like that of the female; length, about $\frac{1}{12}$in.; colour white; pellicle, black, at one end; not carinated.

Some puparia, both male and female, are found slightly curved.

Adult female small, elongated, segmented; length, about $\frac{1}{30}$in., shrivelling at gestation. Cephalic portion compressed, cylindrical. Abdomen somewhat elongated, ending in a number of sharp-pointed, triangular, tooth-like lobes, between which may be made out a few (four?) very minute, roundly-triangular lobes.

Five groups of spinnerets, the three upper groups almost joined in an arch; in the arch, forty to fifty orifices; in the two lower groups, ten to fifteen.

Adult male, brown. Antennæ, ten-jointed; each joint except the first two long and hairy; on the last joint one hair longer than the rest, and ending in a knob. Legs, slender; claw, very thin; digitules, fine hairs.

Habitat—On *Dendrobium*, sp., *Hedycarya*, sp., Hawke's Bay; *Phormium tenax*, *Cordyline australis*, *Astelia cunninghamii*, *Muhlenbeckia*, sp., Wellington; Canterbury; Nelson.

Group II.—LECANIDINÆ.

Female insects flat, convex, or globular; elongated or circular; naked, or covered with waxy, horny, glassy, cottony, or felted secretion forming a covering or test. Adults sometimes apodous and without antennæ. Abdomen in all stages exhibiting a more or less defined cleft, and, above or beside it, on the dorsal surface, two more or less conspicuous, roughly triangular, lobes. Mentum usually monomerous or dimerous.

Male larvæ resembling females. Male pupæ covered with a test of secretion, waxy or glassy. Male adults not greatly differing from Diaspidinæ; abdominal spike usually short and thick.

SUBDIVISIONS AND GENERA.

Subdivision I.

Insects covering themselves with a secretion, composed chiefly of waxy, horny, or glassy matter — LECANODIASPIDÆ.

Test of female horny, partly formed of the second pellicle — LECANOCHITON.

Test of female waxy, with a single fringe of broad segments at the edge — CTENOCHITON.

Test of female glassy, elevated, striated with air-cells — INGLISIA.

Not yet represented in New Zealand.

Test of female waxy, produced into radiating branches — VINSONIA.

Test of female waxy, without fringe or branches — CEROPLASTES.

Test of female waxy, elevated, not striated with air-cells — FAIRMAIRIA.

Tests agglomerated in a waxy mass containing colonies of insects, male and female — CARTERIA.

Test of female absent; tests of males aggregated in a waxy mass — ERICERUS.

Subdivision II.

Female insects naked	LECANIDÆ.
Females propagating without ovisac, arboreal	LECANIUM.
Females constructing ovisac, arboreal	PULVINARIA.

Not yet represented in New Zealand.

Females propagating without ovisac, subterranean, retaining feet and antennæ	LECANOPSIS.
Females propagating without ovisac, subterranean, losing feet and antennæ	ACLERDA.

NOTE.—The genus *Physokermes*, Targioni-Tozzetti, is placed by Signoret (loc. cit., 1874, p. 87) amongst those which here form this subdivision; but there seems to be no sufficient distinction between it and *Lecanium*.

Subdivision III.

Insects covering themselves with secretion of cottony or felted matter	LECANO-COCCIDÆ.
Secretion felted, scarcely apparent on adult female, conspicuous on male pupæ and female of early stages; edge fringed	ERIOCHITON.

Not yet represented in New Zealand.

Secretion felted, appearing only in the last stage, after gestation	SIGNORETIA.
Secretion felted, forming a nearly complete sac on adult female before gestation	LECANODIASPIS.
Secretion felted, forming complete sac on adult female	PHILIPPIA.
Secretion cottony, covering adult female only after gestation	LICHTENSIA.
Secretion cottony, forming a complete sac on adult female before gestation	ERIOPELTIS.

SUBDIVISION I.—LECANODIASPIDÆ, Targioni-Tozzetti.

Female insects exhibiting in all stages the abdominal cleft and lobes. Larvæ free, naked, active. Females after the first metamorphosis constructing over themselves a carapace, shield, or (as called herein) test, of glassy, waxy, or horny secretion. Test apparently homogeneous, really in most cases formed of aggregated segments over the insect, and prolonged beyond the body in a flat single fringe of separate, more or less triangular or quadrate, segments. Fringe not always present or conspicuous in all stages. Males, after first metamorphosis, constructing a test of similar character but varying form.

This subdivision, when first established by Targioni-Tozzetti, included only the four genera *Pollinia, Asterolecanium, Planchonia*, and *Lecanodiaspis*, all of which are apodous in the adult stage of the female. In this work the two first of these are included in a new group, HEMI-COCCIDINÆ; *Planchonia* belongs to the group COCCIDINÆ; *Lecanodiaspis* is left in the present subdivision. For the reasons leading to these changes, and the inclusion in this subdivision of insects retaining the feet in all stages, see N.Z. Transactions, Vol. XI., 1878, p. 207, and Vol. XVI., 1883, pp. 125–128.

Genus: LECANOCHITON, Maskell.

N.Z. Trans., Vol. XIV., 1881, p. 222.

Test of adult female horny in texture, formed partly of secretion, partly of the pellicle of the second stage; abdominal cleft and lobes normal.

30. LECANOCHITON METROSIDERI, Maskell.

N.Z. Trans., Vol. XIV., 1881, p. 222; Vol. XVI., 1883, p. 129.

(Plate VII., Fig. 1.)

Test of adult female brown, hard, horny-looking, convex, slightly elongated, open beneath, loosely attached to twigs by the edges; at the top is the pellicle of the second stage, which is flat, and gives the test the appearance of an overturned basket, of which the pellicle is the foot. Length of test about $\frac{1}{15}$ in. Remains of the thin white test of the second stage may sometimes be seen on the pellicle.

Test of the male small, white, glassy, elongated, convex.

The young insect, extremely minute, naked and active, is flat, oval, brown, or rather reddish, usually found at the tips of young shoots or on leaves. The antennæ have six joints; on the last joint are several hairs, amongst which is one excessively long, slightly knobbed. Foot normal; the joints hairy; upper digitules fine knobbed hairs, lower pair a little broader.

In the second stage the female is scarcely altered: the antennæ and feet remain as before; but there is a test, white, waxy, very thin, covering the dorsal surface, and extending a little beyond the edge in an irregular fringe. On the edge also are a number of protruding spinneret tubes, glassy, white, cylindrical, either curved or straight: a few of these tubes protrude on the surface of the back.

Adult female dark-brown in colour, filling the test; convex above, flat beneath. Rostrum comparatively large; mentum probably monomerous. Antennæ short, thick, atrophied; seven-jointed, but the joints are much confused; on the last joint some hairs. Feet absent. Four rows of rather large spinnerets radiate from the median region of the dorsum to the edge, and along these, on the lower side of the test, are corresponding narrow lines of white cottony secretion.

This insect is viviparous, the young being sheltered awhile by the mother, whose under-side becomes concave during gestation.

Adult male dark-red; length, about $\frac{1}{20}$in. Antennæ of ten joints, of which the two first are very short; the third much longer and expanded at the end; the fourth more than twice as long as the third; the remainder about equal in length to the third, but thicker and rounder, being almost moniliform. All but the first two joints bear hairs. Foot normal; digitules fine hairs.

Habitat—On *Metrosideros robusta* (Rata), Milford Sound; Bluff Harbour. On *M. tomentosa* (Pohutukawa), Auckland.

A peculiar species, easily identifiable by the presence of the second pellicle on the female test.

Genus: CTENOCHITON, Maskell.

N.Z. Trans., Vol. XI., 1878, p. 208.

Test of female waxy, with a single fringe of tooth-like, more or less broad, segments round the edge.

Test of male waxy or glassy, with similar fringe.

The edge of the body, in the second stage of the female, usually presents a wavy appearance, formed by a series of re-entering curves. This is perhaps most conspicuous in *Ct. perforatus*.

The antennæ of the adult female have six or seven joints. It is often very difficult to determine the number, as the third joint exhibits frequently a shallow circular depression or ring which may easily be mistaken for a division.

During gestation the female, which at first fills the test, shrivels up at one end into a shapeless mass requiring maceration in potash to restore the original form for examination. The test thus becomes almost filled with eggs.

The presence of the fringe, which is noticeable in every species of this genus at some time or other, distinguishes it from *Ceroplastes* and *Vinsonia*. In *Ctenochiton viridis* the test is not to be made out in the adult stage but is clearly visible in the second stage of the female, and it is present, with the characteristic fringe, for the male.

31. CTENOCHITON DEPRESSUS, Maskell.

N.Z. Trans., Vol. XVI., 1883, p. 132.

(Plate VII., Fig. 2.)

Test of adult female flat, nearly circular, thin, waxy, greyish-coloured; fringe inconspicuous or sometimes absent. No perforations or rows of air-cells. Diameter, about $\frac{1}{4}$in.

The fringe is more conspicuous in the test of the second stage.

Test of male elongated, narrow, flat beneath, slightly convex above, white, glassy, thin and brittle, with a conspicuous fringe of which the segments are truncato-triangular. The test is divided into tessellations, the median row of which is quadrangular, with two series of pentagonal divisions between it and the fringe. Near the abdominal extremity a transverse narrow slit cuts the test in two, leaving a small segment at the extreme end apparently separate. Length of the test about $\frac{1}{14}$in.

Adult female filling the test, but shrivelling after gestation; colour brownish or grey. Antennæ of six joints, the third being the longest, and, as is commonly the case in the genus, often appearing like two. On the last joint a few long hairs. Foot normal; the upper digitules are fine hairs, the lower pair only a little broader. Anal ring and lobes normal.

In the second stage the usual wavy edge is conspicuous; the test is thin, glassy, with normal fringe. Antennæ and feet normal. The insect is somewhat thick, with yellowish colour.

Young insect normal.

Adult male yellowish-red in colour, about $\frac{1}{25}$in. in length, exclusive of the wings. General form normal. Antennæ long, with ten joints, all long and equal, except the two first, which are very short: all the joints have several hairs. Legs normal, but the tibiæ are very long and slender, and only a little thickened at the tip; tarsi somewhat thick; digitules fine hairs. Abdominal spike long, and very slightly curved.

Habitat—On *Plagianthus*, *Cyathea*, &c.; Hawke's Bay.

This insect resembles, to the naked eye, somewhat nearly *Ctenochiton perforatus*, but the female differs in the absence of the curious perforations in the test of that species, and in the shorter and thicker antennæ, with also more long hairs on the last joint. The test of the male is also different.

32. CTENOCHITON ELÆOCARPI, Maskell.

N.Z. Trans., Vol. XVII., 1884, p. 26.

(Plate VII., Fig. 3.)

Test of adult female oval, nearly circular, black in colour, divided into hexagonal and pentagonal segments which are not conspicuous, and of which the median series forms a very slightly elevated ridge somewhat lighter in colour. The test is only slightly convex. The fringe is very long and conspicuous, the segments toothlike. Diameter of test, exclusive of the fringe, reaches $\frac{1}{6}$in.

Test of male unknown.

Test of second stage of female white, waxy, not homogeneous, but built up of a number of loosely-aggregated tubuliform plates, somewhat resembling those of the genus *Orthezia*, Bosc. The fringe of this test is longer than in that of the adult, the teeth curling in different ways. Length of test and fringe, sometimes $\frac{1}{8}$in.

The adult female fills the test, shrivelling at gestation. Colour black. Antennæ somewhat long, of seven joints; a few hairs on the last joint. Foot normal; upper digitules strong and thick, lower pair very broad. On the skin are a number of large oval spots which appear to be the orifices of spinneret tubes.

Female of the second stage wanting the usual wavy edge of the genus. Round the edge of the body is a row of sharp conical spines set closely together. Antennae of six somewhat confused joints. Feet normal.

Adult male unknown.

Habitat—On *Elæocarpus dentatus* (hinau), Wellington.

The large size, black colour, and very conspicuous fringe of the adult female test, and the white loose test of the second stage, distinguish this species, especially from *Ct. fuscus*.

33. CTENOCHITON ELONGATUS, Maskell.

N.Z. Trans., Vol. XI., 1878, p. 212.

(Plate VII., Fig. 4.)

Test of adult female elongated, narrow, convex. Length sometimes reaching $\frac{1}{4}$in., width about $\frac{1}{16}$in. Fringe not always conspicuous; the segments quadrate outwardly. Test divided into quadrangular divisions. Colour whitish, but often blackened by fungoid growths.

Test of male unknown.

Adult female filling the test, shrivelling at gestation. Edge of the body wavy. Stigmatic spines long and conspicuous. Antennæ seven-jointed. Feet normal; lower digitules absent (?).

Adult male unknown.

Habitat—On *Geniostoma ligustrifolium*, Auckland; on *Dendrobium* sp., Hawke's Bay; on *Earina* sp., Wellington.

Easily recognized by the great length and narrowness of the female test.

34. CTENOCHITON FLAVUS, Maskell.

N.Z. Trans., Vol. XVI., 1883, p. 130; Vol. XVII., 1884, p. 26.

(Plate VII., Fig. 5.)

Female test golden, waxy, flat beneath, convex above; outline circular or slightly elliptical, with a fringe of broadly triangular segments round the edge. Apex of the test an irregular elongated mass of wax, the remainder divided into two concentric series of plates, the inner series pentagonal with sharp angles, the outer pentagonal with rounded angles and with the outer side forming the base of the segments of the fringe. The inner series forms often irregular lumps of wax. Diameter of test sometimes reaching $\frac{1}{5}$in. The colour is often hidden by black fungoid growths.

Test of male much narrower than that of the female, having an irregularly rectangular edge with deep curvilinear depressions. It is glassy, white and shining, flat beneath and elevated above, and marked with numerous horizontal striae. The upper central portion is sometimes flat, sometimes an irregular mass of the glassy secretion. On the lower side there is often a plate of secretion, so that the pupa is almost entirely enclosed.

The adult female fills the test, shrivelling up after gestation: it is consequently flat beneath, convex above, with general outline of Lecanidinae. The spiracular spines are very long and conspicuous; from their base a double row of minute circular spinneret orifices runs as far as the spiracle, with two or three outlying ones at the base of the spine; and a single row of the same kind of orifices runs across the body to the spiracle on the other side. Along the edge of the body there is a series of conical sharp spines; and scattered all over are many tubular projecting spinnerets. The abdominal cleft is deep, and the two lobes are conspicuous on the dorsal side; these lobes are not smooth, but irregular, and each bears at the end three or four strong spines. The antennae have six joints; but the third joint often looks like two on account of the false division or depressed ring: the last joint has several long hairs. Feet normal; the upper digitules fine long hairs, the lower pair very broad. The anal ring bears a number of long hairs, of which eight seem to be conspicuous. The colour of the insect is a golden brown; diameter averaging $\frac{1}{12}$ in.

The second stage of the female is normal of the genus, showing the wavy outline, somewhat strongly marked in many specimens, but not conspicuous in others. The spiracular spines are prominent, and a row of conical spines runs round the edge of the body, as in the adult. The test is at first very thin and brittle, and with a fringe of broad, shallow segments; but afterwards becomes thicker, and in the end, before the change to the final stage, it approaches almost the form of the waxy test of an adult *Ceroplastes*.

The young insect is normal.

The adult male is normal of the genus. The legs are very long and slender; the four digitules are fine hairs. At the extremity of the tibia there is a strong spine. Abdominal spike, or sheath of the penis, slightly curved, with a seta on each side of its basal tubercle. Antennæ of ten joints; the first two very short, the rest longer and equal. On the last joint are several long hairs, of which three are knobbed.

Habitat—On *Brachyglottis repanda, Panax arboreum, Leptospermum scoparium* (manuka), *Elæocarpus dentatus* (hinau); Wellington.

The species is distinguishable by the shape and colour of the test and the arrangement of the spinnerets, in the adult female. The tests of the second stage may sometimes be taken for adult *Ceroplastes rusci*, Linn.; but can be easily distinguished on examination of the enclosed insect.

35. CTENOCHITON FUSCUS, Maskell.

N.Z. Trans., Vol. XVI., 1883, p. 131.

(Plate VII., Fig. 6.)

Test of the adult female elliptical in outline, flat below, convex above, the elevation being greater than usual; almost black in colour, composed of a thin dark waxy secretion. The fringe is conspicuous, and has the appearance of teeth, the segments being sharply triangular and set closely together. Length of test sometimes nearly $\frac{1}{4}$in., breadth $\frac{1}{7}$in., height $\frac{1}{16}$in. Inside of the test whitish.

Test of male glassy, white, elongated, slightly convex.

Adult female filling the test, shrivelling at gestation. Antennæ short, probably seven-jointed, but the joints are confused; on the last joint several hairs. Foot having the tibia expanded at the extremity; upper digitules strong and thick, lower pair ending in conspicuously broad plates. On the edge of the body a row of conical spines. Colour almost black.

In the second stage the female is less wavy in outline than in other species of the genus, and in its later period is somewhat thick, with the edges turned inwards. Feet normal; digitules fine. Antennæ short and thick, with six joints, of which the third and fourth are the longest; on the last joint some long hairs. The abdominal lobes are irregularly triangular.

Young insect normal.

Adult male unknown.

Habitat—On *Brachyglottis repanda, Panax arboreum*; Port Hills, Canterbury (Dry Bush).

Distinguished by its large size, great convexity, and black

colour, which is not due to fungoid growths, although these, as usual, accompany it.

36. CTENOCHITON HYMENANTHERÆ, Maskell.

N.Z. Trans., Vol. XVII., 1884, p. 25.
(Plate VIII., Fig. 1.)

Test of adult female waxy, circular, convex, dirty-white, yellow, or brownish, formed of a number of hexagonal or octagonal segments, which are also convex, giving it a rough appearance. Fringe not very conspicuous. Diameter of test, about $\frac{1}{12}$in.

Test of male glassy, dirty-white, oval, segmented, slightly convex, segments of fringe small. Length, about $\frac{1}{16}$in.

Adult female yellowish-brown, filling the test. Antennæ of six joints, of which both the second and third seem sometimes double. Foot normal; upper digitules long fine hairs, lower pair broad. The spiracular spines are strong and conspicuous. The skin is divided into segments corresponding with those of the test, the divisions being marked by lines of spinneret orifices which are small and simple.

In the second stage the usual wavy edge of the genus is not generally apparent.

Adult male somewhat thick and short. Antennæ of nine joints, the first short and thick, the remainder long and nearly equal; each joint after the first has many nodosities, from which spring longish hairs. Foot long and slender, especially the tibia. Digitules fine hairs. Thoracic band inconspicuous. Abdominal spike short and blunt.

This species is usually accompanied by a great quantity of very black fungus covering and rendering unsightly the whole plant on which it lives.

Habitat — On *Hymenanthera crassifolia*, Evans Bay, Wellington.

This insect seems to be intermediate between *C. piperis* and *C. depressus*, differing from both in the rugose female test and the distribution of the spinneret orifices.

37. CTENOCHITON PERFORATUS, Maskell.

N.Z. Trans., Vol. XI., 1878, p. 280; Vol. XVI., 1883, p. 130.

(Plate VIII., Fig. 2.)

Test of adult female white, waxy, circular, nearly flat, brittle, thin except at the edge. Fringe thin, segments broadly triangular. Diameter nearly $\frac{1}{8}$in. The test is divided by narrow lines of minute spots, corresponding to the spinneret orifices of the insect, into rows of pentagonal or hexagonal segments. The interior segments are only dotted, but the exterior row exhibits curvilinear series of small perforations or air-cells arranged in slightly radiating rows, which extend also to the corresponding segments of the fringe.

The test of the second stage of the female is very thin and filmy, waxy, flat, slightly elongated; the fringe as in the adult; but there are no perforations or air-cells. Length, about $\frac{1}{10}$in.

Test of male waxy, thin, slightly elongated, rather convex; length, about $\frac{1}{15}$in. Fringe and air-cells as in the female test. At the abdominal end is a joint or hinge separating the last segment.

Adult female filling the test, shrivelling at gestation; colour greyish or greenish-white. General appearance somewhat leathery. Rows of minute oval spinnerets run round the edge and across the body, corresponding with the divisions of the test. Antennæ of six joints, the third joint being the longest, and seeming double on account of the depressed ring: on the last joint some hairs. Feet normal, with somewhat thick coxæ and femora: upper digitules long; lower pair narrow. Anal ring with eight hairs. A row of scattered small hairs runs round the edge of the body.

Female of second stage very thin and transparent, seeming like a bluish-green film. The wavy edge is conspicuous. Antennæ of six joints.

Adult male yellowish. Antennæ nine-jointed, every joint except the first bearing several hairs. Feet normal; upper digitules not long, lower pair fine hairs. Thoracic band conspicuous and long. Abdominal spike short.

Habitat—On *Pittosporum eugenioide*, *P. tenuifolium*, *Panax arboreum*, *Coprosma lucida*, *Rubus*, &c., Riccarton Bush, Christchurch; Nelson; Dunedin; Wellington.

38. CTENOCHITON PIPERIS, Maskell.

N.Z. Trans., Vol. XIV., 1881, p. 218; Vol. XVII., 1884, p. 25.

(Plate VIII., Fig. 3.)

Test of adult female circular, convex, regularly tessellated in hexagonal segments; fringe not very regular, sometimes almost or quite absent; waxy, somewhat thick; diameter, about $\frac{1}{12}$in. Colour variable; outer parts white, yellow, or greenish-white, central segments purplish.

Test of second stage of female waxy, very thin, flat, with fringe of broadly triangular segments: no air-cells. Length, about $\frac{1}{30}$in.

Test of male elongated, convex, glassy, segmented. Length, about $\frac{1}{20}$in. Slightly coloured like that of the female.

Adult female filling the test. Colour corresponding with that of the test. At gestation the under-side becomes hollow, and the young are sheltered beneath it for awhile. Antennae seven-jointed; on the last joint several hairs. Feet normal; lower digitules rather broad. Round the dorsal surface, half-way between the centre and the edge, is a row of swellings* or tubercles.

Female of second stage elongated, the cephalic portion narrower than the abdomen; flat, thin; edge wavy; stigmatic spines rather stout; a few small spines on the edge. Antennae six-jointed; feet normal.

Young insect of normal form, but with numerous minute wrinkles on the edge of the body.

Adult male greenish-yellow; antennae nine-jointed. Foot normal, with somewhat thick tarsus. Penis ending in a round, somewhat large knob.

Habitat—On *Piper excelsum* (kawakawa); Auckland, Hawke's Bay, Wellington.

The regularly-circular form and coloured segments of the female test, and the tubercles of the dorsum, distinguish this species.

---

* Possibly spiracular.

39. CTENOCHITON VIRIDIS, Maskell.

N.Z. Trans., Vol. XI., 1878, p. 211; Vol. XVII., 1884, p. 24.

(Plate IX., Fig. 1.)

Test of adult female absent or fragmentary; very thin, waxy, white, divided into pentagonal or hexagonal segments, each of which exhibits numerous concentric wavy curves, crossed by straight lines radiating from the centre;* the segments are separated by double lines of minute spots, corresponding to the spinneret orifices of the insect. No air-cells. Fringe seldom visible.

Test of female of second stage very thin, white, waxy, flat, divided into segments with concentric curves and radiating lines as in the adult. Fringe of broad segments. No air-cells. Length, about $\frac{1}{15}$in.

Test of male glassy, white, elongated, slightly convex, divided into segments similarly marked to those of the female. Abdominal segments separated from the test by a transverse line, or hinge. Length, about $\frac{1}{6}$in.

Adult female bright-green in colour, thick, elongated or pyriform, the cephalic portion somewhat acuminate; length sometimes reaching $\frac{1}{2}$in. It produces a conspicuous depression in the leaf, in which the body is partially buried. Antennæ of six joints (apparently seven, but the third joint shows the false division or depressed ring), often atrophied. Feet normal, coxæ and femora thick. After gestation the insect frequently becomes brown, covered with a mass of white mealy or felted secretion.

Female of the second stage thin, filmy, translucent; flat, elongated, with wavy outline; colour green; length, about $\frac{1}{20}$in. Antennæ of six joints.

Adult male greenish-yellow; length, about $\frac{1}{15}$in. Antennæ of nine joints. Feet normal; digitules absent. A strong spine at the extremity of the tibia.

Habitat—On *Panax arboreum*, *Coprosma lucida*, *Hedycarya dentata*, *Atherosperma Noræ-Zælundiæ*, *Rubus australis*; Canterbury, Otago, Wellington, Nelson, Auckland, Hawke's Bay.

This is probably the largest known species of the Lecanidinæ. Its size and bright-green colour in the adult state clearly dis-

---

* The radiating lines and concentric curves of these segments are usually somewhat conspicuous, at least on the adult female, and serve to distinguish the test from that of *C. elongatus*, which otherwise resembles it.

tinguish it. The female of the second stage resembles nearly that of *Ct. perforatus*, but is somewhat thicker, and the markings of the segments of the test are different.

Genus: INGLISIA, Maskell.

N.Z. Trans., Vol. XI., 1878, p. 213.

Test of female glassy, elevated, striated with radiating rows of air-cells. Fringe not always present in the adult stage.

In the genus *Fairmairia*, Signoret, there is also an elevated test, but it is waxy, and exhibits no air-cells, and has no fringe in any stage.

40. INGLISIA LEPTOSPERMI, Maskell.

N.Z. Trans., Vol. XIV., 1881, p. 220; Vol. XVII., 1884, p. 27.

(Plate IX., Fig 2.)

Test of adult female white, glassy or waxy, elongated, convex above, flat and open beneath, formed of several agglutinated segments, each segment more or less convex or conical, median segments usually five in number; at the edge an irregular fringe, but the fringe is often absent. Average length of test, $\frac{1}{10}$in. The marginal segments sometimes assume the form of small cones, as if a number of secondary tests were attached to the principal one. All the segments are marked with striæ radiating from the apex of each: the striæ, which are composed of air-cells, widen from the apex to the base.

Test of the male white, glassy, elongated, convex, not unlike that of the female, but with a longer fringe; it has also its posterior segment divided from the rest by a transverse slit or hinge; average length, about $\frac{1}{15}$in.

Adult female filling the test, shrivelling at gestation; colour brown; abdominal lobes yellow, conspicuous. The flat undersurface is smooth; the dorsum divided by large corrugations, each segment corresponding to one in the test. Antennæ of seven joints, of which the third is the longest, the fourth, fifth, and sixth the shortest; a few hairs, especially on the last joint. Feet normal; the tibia is somewhat thin, and has one spine or hair at its tip. Digitules normal; upper pair long knobbed hairs, lower pair very broad.

The female in the second stage is also convex above, flat below, but is less thick than the adult, and has not the corrugations. General form elongated-oval; the abdominal lobes are not, as usual, smooth, but approach by irregularity the anal tubercles of the Coccidinæ, and like them bear a few hairs. The anal ring has eight hairs. Antennæ of six joints. Feet normal; digitules as in adult. On the skin are several scattered, circular, very minute spinnerets; the stigmatic spines are long and conspicuous, and along the edge runs a row of conical hairs or spines.

Adult male yellowish-green in colour, the body slender and tapering. From the abdomen spring two very long white cottony setæ, one on each side of the spike, which is straight and short. Antennæ of ten joints; the first two short, the rest long, thin, and hairy. Of these, the seventh, eighth, and ninth are the shortest; on the last joint three long knobbed hairs. Feet slender, hairy; digitules normal. Thoracic band inconspicuous.

Habitat—On *Leptospermum scoparium* (manuka); Christchurch, Kaiapoi, Wellington, Auckland. It affects the twigs of the plant, and not the leaves.

41. INGLISIA ORNATA, Maskell.

N.Z. Trans., Vol. XVII., 1884, p. 27.

(Plate X., Fig. 1.)

Test of adult female reddish-brown, the base more or less oval, the rest elevated in a cone and ending in a prominence standing up like a more or less sharp horn; sometimes there are two of these horns. The test is formed of a number of polygonal segments, each slightly elevated, and all are marked with the radiating striæ peculiar to the genus. There is a fringe of sharply triangular segments, also striated. Average length of test, about $\frac{1}{8}$in., but specimens attain a length of $\frac{1}{4}$in.; height, about $\frac{1}{16}$in.

Test of second stage generally resembling that of the adult, but smaller and less conical, and more tinged with green; and at the edge a number of short spinneret tubes may be seen protruding.

Test of the male elongated-oval, convex, but wanting the prominent horn of the female, glassy, white tinged with yellowish-brown, composed of segments marked with conspicuous striae. Length, $\frac{1}{12}$in. Fringe often present, but irregular; often absent.

The adult female fills the test, shrivelling after gestation. It exhibits the horn, or two horns, as in the test. Antennae of seven joints, the third joint showing the false division noted in other species of Lecanodiaspidae. Feet normal; upper digitules strong and thick, lower pair rather broad. Along the edge of the body is a row of sharp lanceolate spines set closely together, and the spiracular spines are long and conspicuous. A double or triple row of minute circular spinnerets marks the divisions corresponding to the segments of the test. Colour of the insect greenish, turning brown after gestation. The abdominal lobes are brown.

In the second stage the female resembles generally the adult; but the antennae have six joints, and amongst the marginal spines are some very much larger than the rest.

The young larva is flat and oval, and at the margin shows a fringe of long glassy pointed tubes, springing from the marginal spines.

The adult male is about $\frac{1}{10}$in. in length (exclusive of the wings), brownish or reddish-yellow in colour, the wings hyaline and iridescent, with red nervures. Antennae of ten joints, on the last of which are, amongst others, three long knobbed hairs. Foot with a spine at the extremity of the tibia; digitules fine hairs. At each side of the abdominal spike springs a strong seta, from which extends a white cottony pencil, as long as the body of the insect. The penis is a long soft cylindrical tube covered with minute recurved spines. Thoracic band short and narrow.

Habitat—On *Elaeocarpus dentatus* (hinau), *Leptospermum scoparium* (manuka); Wellington.*

This is a handsome species: the colour and the horns of the test are clear distinctions.

---

* The male pupae may be found not unfrequently on other plants, such as *Coprosma*, *Pittosporum*, &c.

42. INGLISIA PATELLA, Maskell.
N.Z. Trans., Vol. XI., 1878, p. 213; Vol. XIV., 1881, p. 219.
(Plate X., Fig. 2.)

Test of adult female conical or limpet-shaped, white, glassy, slightly elongated, striated with rows of air-cells radiating from the apex, and increasing in size to the edge. Length of test, about $\frac{1}{15}$in.; height, about $\frac{1}{40}$in. The edge is usually very wavy.

Test of male similar, but a little smaller and more elongated.

Adult female filling the test, shrivelling at gestation. Colour greenish-yellow. Edge wavy, corresponding to the curves of the test. Antennæ very short, six-jointed; on the last three joints some hairs. Feet normal; upper digitules very long, lower pair narrow. On the edge of the body is a row of spines, of which each alternate one is conical, the remainder club-shaped; the edge seems double or ribbon-like, and inside it is a row of spinnerets, beyond which is a narrow line of short regular curves. Abdominal cleft wide and circular above, the extremities almost meeting. Anal ring with eight long hairs.

Female of second stage flattish, elongated, wavy-edged; not exhibiting alternate conical and clavate spines.

Young larva naked, flat, active; round the edge a row of clavate spines, but no conical spines. Length, about $\frac{1}{50}$in.

Adult male greenish-yellow; length, $\frac{1}{24}$in. Antennæ of ten joints, all except the first bearing hairs. Foot normal; digitules fine hairs. Abdominal spike about half the length of the abdomen, with two long white setæ springing from the basal tubercle.

Habitat—On *Coprosma lucida*, Riccarton Bush, Christchurch; *Drimys colorata*, (plentifully) on hill above Lyttelton; *Atherosperma*, Wellington.

A very pretty little species, clearly distinguished by the form of the test and the alternate spines of the adult female.

Subdivision II.—LECANIDÆ.

Female insects naked in all stages; form variable; apodous in adult stage, or retaining the feet; viviparous or oviparous, with or without attached ovisac; abdominal cleft and lobes always present. Male pupæ in some cases covered with waxy secretion.

Genus: LECANIUM, Illiger.

Females naked, flat or convex; viviparous or oviparous; propagating without ovisac; arboreal.

Dr. Signoret (loc. cit., 1873, p. 396) divides the genus into six series, as follows:—

(1.) Species flat, usually viviparous; example, *L. hesperidum.*
(2.) Species more or less convex, elongated; example, *L. persicæ.*
(3.) Species more or less globular, the skin tessellated; example, *L. aceris.*
(4.) Species more or less globular, the skin perforated with oval markings; example, *L. hemisphæricum.*
(5.) Species rugose, with dorsal keels; example, *L. oleæ.*
(6.) Species globular, with cleft beneath for attachment to twigs; example, *L. emerici.*

The following are the only species reported as yet in New Zealand; but the genus is so widely spread and the species are so numerous that doubtless many others will hereafter occur in this country.

13. LECANIUM DEPRESSUM, Targioni-Tozzetti, Catal. (1868), 37, 8; Stud. sul Coccin., 29.
Maskell, N.Z. Trans., Vol. XI., 1878, p. 206.

(Plate XI., Fig. 1.)

Adult female elongated, somewhat acuminate at the cephalic end, slightly convex; reddish-brown; skin marked with two dorsal keels and numerous irregular tessellations, finely punctate. Antennæ of eight joints; on the first two and the last three joints some hairs. Feet normal, rather long; one of the lower digitules is larger than the other. Length of insect, about $\frac{1}{10}$ in.

Male unknown.

Habitat in New Zealand—On plants in greenhouses; Christchurch, Wellington. In Europe, on *Ficus*, in hothouses.

This insect belongs to Signoret's fifth series.

44. LECANIUM HEMISPHÆRICUM, Targioni-Tozzetti, Stud. sui Coccin., 27.

Chermes filicum, Boisduval (1867), 336; Targioni, Catal., 1868, 38, 17.

Maskell, N.Z. Trans., Vol. XVII., 1884, p. 29.

(Plate XI., Fig. 2.)

Adult female hemispherical, with broad flattened edges; reddish-brown; diameter, about $\frac{1}{15}$ in. Skin regularly marked with oval perforations; no keels. Antennæ of eight joints. Feet normal, thin. Anal ring with eight hairs.

Male unknown.

Habitat in New Zealand—On Camellia, Hutt Valley, Wellington; in Europe, on *Dracæna australis*; in America, on various greenhouse plants (Comstock). Query—Does the European habitat denote an Australian origin?

This insect belongs to Signoret's fourth series.

45. LECANIUM HESPERIDUM, Linnæus.

L. hesperidum, Linnæus, Syst. Nat., 1735, II., 739, 1; Faun. Succ., 1746, 1015.

Coccus hesperidum, various authors.

Calymnatus hesperidum, Costa, Nuov. Osserv., 1835?

Calypticus hesperidum, Costa, Faun. Ins. Nap. Gall-insect., 1837, 8, 1; Lubbock, Proc. Roy. Soc., IX., 480; Beck, Trans. Roy. Micr. Soc., 1861, 47, &c.

Maskell, N.Z. Trans., Vol. XI., 1878, p. 205.

The Holly and Ivy Scale.

(Plate XI., Fig. 3.)

Adult female naked; yellow, brown, or reddish; flat or slightly convex; elongated; skin smooth, sparsely punctate; length averaging $\frac{1}{10}$ in., but specimens reach sometimes $\frac{1}{5}$ in. Antennæ of seven joints; a few hairs on most, but the seventh has several. Abdominal cleft and lobes normal. Feet normal. On the edge of a body a row of small hairs, not set closely together. Viviparous; at gestation the under-side becomes concave, forming a shelter for the young; and this cavity is often of a blood-red colour. On the under-side may be seen sometimes four cottony trails starting from the region of the four stigmata.

Young larva reddish-brown; oval, flat; antennæ of six joints. From the abdominal lobes spring two long setæ.

Male unknown.

Habitat in New Zealand—Everywhere, on ivy, holly, camellia, orange, laurel, myrtle, box, and many other plants out of doors or in greenhouses. In Europe, chiefly on ivy and oranges, but frequently on other plants. In America on many plants.

This is the commonest of the Lecanidæ in this country; it may be distinguished from *L. mori* (below) by its flatness and sparse punctuation.

This insect belongs to Signoret's first series.

16. LECANIUM HIBERNACULORUM, Targioni-Tozzetti, Catal. (1868), 37, 9.

Chermes hibernaculorum, Boisduval, Ent. Hort., 1867, 337.

Maskell, N.Z. Trans., Vol. XI., 1878, p. 207.

Adult female nearly globular; naked; reddish-brown; diameter about $\frac{1}{4}$in.; at gestation the body becomes simply an inverted bag covering the eggs and young. The insect appears to be partly oviparous, partly viviparous. Antennæ of eight joints. Feet normal. Skin pretty regularly marked with oval perforations.

Male unknown.

Habitat in New Zealand—On various greenhouse plants, Christchurch. In Europe on *Brexia*, *Phajus*, &c.

The insect belongs to Signoret's fourth series, and may perhaps be only a large variety of *L. hemisphæricum*.

17. LECANIUM MACULATUM, Signoret, Ann. de la Soc. Ent. de France, 1873, p. 400.

Maskell, N.Z. Trans., Vol. XI, 1878, p. 207.

Adult female naked; flat, elongated; dorsal skin marked with a median row of rather large oval spots reaching from the abdominal cleft to the region of the rostrum. Length, about $\frac{1}{5}$in. Colour yellowish-brown. Antennæ of seven joints. Feet normal.

Male unknown.

Habitat in New Zealand—On *Bavardia*, in hothouses, Christchurch. In Europe on ivy.

This insect belongs to Signoret's first series: the dorsal spots distinguish it from *L. hesperidum*.

18. LECANIUM MORI, Signoret, Ann. de la Soc. Ent. de France, 1873, p. 407.

    Maskell, N.Z. Trans., Vol. XVII., 1884, p. 29.
        (Plate XI., Fig. 4.)

Adult female naked; elongated, convex; reddish; length, about $\frac{1}{8}$in. Skin smooth, without spots, tessellations, or keels. Antennæ of seven joints. Feet normal.

Adult male unknown: pupa covered by a white, elongated, segmented, glassy test.

Habitat in New Zealand—On *Alsophila Colensoi* and other ferns, Botanical Gardens, Wellington. In Europe on mulberry, &c.

The insect belongs to Signoret's second series.

19. LECANIUM OLEÆ, Bernard.

    Chermes oleæ, Bernard, Mem. d'Hist. Nat. Acad., 1872, 108.
    L. oleæ, Signoret, loc. cit., 1873, p. 440.
    Maskell, N.Z. Trans., Vol. XVII., 1884, p. 28.
        The " *Black Scale*."
        (Plate XI., Fig 5.)

Adult female naked; semi-globular; dark-brown, sometimes almost black. Skin marked by one longitudinal and two transverse keels, not very conspicuous. Diameter, about $\frac{1}{15}$in. Antennæ of eight joints. Feet normal. Anal ring with six hairs.

Young insect flat, elongated, reddish-brown. The keels are more conspicuous than in the adult. Skin marked with numerous oval perforations.

Male unknown.

Habitat in New Zealand—On camellia, *Cassinia leptophylla* (tauhinu), and other plants, Wellington; Hawke's Bay; on various native trees, Whangarei. In Europe on olive. In America (where it is called the "black scale") on oranges and very many other plants.

This insect belongs to Signoret's fifth series.

A European species, *L. cycadis*, Boisduval, is said by Dr. Signoret to closely resemble *L. oleæ*, the only difference apparently being the possession of nine-jointed antennæ. This character is so exceptional in the genus that it perhaps may be but doubtful.

Genus: PULVINARIA, Targioni-Tozzetti.

Female insects naked, arboreal, constructing an ovisac. Male pupæ in cottony or waxy sacs.

50. PULVINARIA CAMELLICOLA, Signoret, Ann. de la Soc. Ent. de France, 1873, p. 32.

Maskell, N.Z. Trans., Vol. XI., 1878, p. 207.

(Plate XII., Fig. 1.)

Adult female yellowish- or reddish-brown, naked, slightly convex, elongated; skin smooth, with puncta; length variable, from about ⅕in. to ¼in. Antennæ (according to Signoret) with sometimes six, sometimes seven, joints. Abdominal cleft and lobes normal. The insect excretes a narrow, white, cylindrical cottony ovisac, which is conspicuous on the leaf of the plant, and the brown body of the female can be seen at one end of it. The eggs in this ovisac are numerous, perhaps some hundreds.

Larva and second stage of female flat, oval, yellowish-brown.

Male pupa covered with a waxy, elongated test as in the genus *Ctenochiton*, but there is no fringe and the segments of the test are not conspicuous; the test is oval and convex.

Adult male yellowish-grey, the head rounded, with an anterior protuberance. Two dorsal and two ventral eyes, and two ocelli. Antennæ of ten joints, all hairy. Feet exhibiting only two digitules, the upper pair. Abdominal spike short, with two longish setæ on each side, each pair of which are covered with cotton which is produced into a long white conspicuous cauda.

Habitat—On camellia. In the South, chiefly in greenhouses. In the Hutt Valley, Wellington, camellias in the open air are much subject to it.

The female of this species is not unlike *Lecanium hesperidum*, but the formation of the white ovisac is a clearly distinguishing character. In late summer the female often drops off to the ground, leaving only the ovisac observable on the leaf.

Subdivision III.—LECANO-COCCIDÆ, Maskell.

N.Z. Trans., Vol. XVI., 1883, p. 128.

Female insects covering themselves with a secretion of cottony or felted matter, forming more or less complete sacs. Male insects (where known) covered with similar secretion.

Genus: ERIOCHITON, Maskell.

Secretion white, felted, formed of threads issuing from prominent spiny spinnerets; inconspicuous or absent on adult female, thicker on male pupa. Abdominal cleft and lobes present in all stages of female.*

[This genus contains the insect named hitherto *Ctenochiton spinosus*.]

51. Eriochiton hispidus, Maskell.

N.Z. Trans., Vol. XIX., 1886, p. 47.

(Plate XIII., Fig. 1.)

Secretion of female white, thin, felted, formed of thin threads excreted from the numerous prominent spiny spinnerets, the threads becoming more or less matted over the dorsal surface. At the edge each thread corresponds to a spine, but has not the feathery form exhibited in the next species, *E. spinosus*, being more tubular. On the adult female the covering is often not to be detected, or presents only fragmentary portions; it is best observed on the female of the second stage.

Secretion of the male pupa white, felted, thick, covering the insect all over, and exhibiting at the edge a small fringe. At first sight the test, being obscurely segmented, presents somewhat the appearance of a Dactylopid. Length of the felted mass, about $\frac{1}{16}$ in.

Larva normal, flat, elliptical, active, exhibiting the usual abdominal cleft and lobes. Dorsal surface covered with spines, excreting a thin white mass of tubes and a tubular fringe.

---

* In both of the species here described the adult female has the tibiæ shorter than the tarsi. The author has hesitated to found a generic character on it until the discovery of other species; the character is quite exceptional in the family, occurring (besides) only in some species of *Acanthococcidæ*. In all others a tibia shorter than the tarsus would indicate an immature insect.

Female of the second stage more or less elliptical, slightly convex, brown in colour beneath the thin white felted covering, which usually presents a segmented appearance, due to the transverse rows of prominent spinnerets. Body covered thickly on the dorsal surface with spines, which are subcylindrical, the ends rounded, springing from tubercular bases. On the ventral surface many smaller spiny hairs. Abdominal cleft normal, the lobes large. Mentum probably monomerous: the tip bears several hairs. Antennae of six somewhat hairy joints. Feet with rather large femora; the lower digitules are fine hairs. Anogenital ring bearing numerous hairs.

Adult female elliptical, convex, hollow beneath, brown in colour, usually affecting the twigs and branches of the plant in preference to the leaves. Apparently naked, but on close inspection found to retain at least portions of the thin felted covering. Dorsal surface covered with great numbers of spines similar to those of the second stage; ventral surface with many small spiny hairs. Antennae of seven joints. The feet have large coxae and femora; the tibia is only about half as long as the tarsus; the lower digitules are only fine hairs.

Adult male of normal form of *Lecanidinae*: colour brown. On the head are six visual organs: two dorsal eyes, two ventral, and two ocelli. Antennae reddish, ten-jointed; the second joint a good deal thicker than the rest, the second, third, and fourth joints the longest; the last three moniliform; all the joints hairy. On the five last joints are several hairs with knobbed extremities. Feet slender, hairy; digitules fine hairs. Abdominal spike short and rather broad. On each side of the base of the spike is a tubercle bearing a pair of longish setae; each pair of setae becomes enclosed in a long white cottony thread, and the two threads form conspicuous "tails," as is common with most males of the Coccid family.

Habitat—On *Olearia Haastii*, Botanical Gardens, Wellington. This is an alpine plant cultivated in the Gardens, and the insect probably came with it from the mountains.

This species is distinguished from *E. spinosus* by the great number of spiny spinnerets on the dorsum of the female and by the tubular character of the fringe.

The curious and exceptional character of a tibia shorter than the tarsus in the adult female, as observed above, is found only in this genus and some *Acanthococcidae*.

52. ERIOCHITON SPINOSUS, Maskell.
Ctenochiton spinosus, Maskell, N.Z. Trans., Vol. XI., 1878, p. 212; Vol. XII., 1879, p. 292; Vol. XIV., 1881, p. 218; Vol. XVII., 1884, p. 25.
(Plate XIII., Fig. 2.)

Test of female white, thin, formed of felted threads excreted from spiny spinnerets; inconspicuous at all stages, and often absent on the adult, but distinguishable on the larva and the second stage. The excreting spinnerets are almost all at the edge of the body, and the fringe is formed of featherlike segments, each segment corresponding to a spine.

Test of male white, thick, felted, oval, and convex; length, about $\frac{1}{16}$ in. Fringe as in the female.

Adult female dark-brown, sometimes almost black; slightly elongated, convex, affecting almost altogether the twigs and branches in preference to the leaves. Average length, about $\frac{1}{9}$ in. Antennæ of seven joints: on the last joint some hairs. Feet with the tibia about half as long as the tarsus (see note above, under the genus). On the edge of the body is a row of conspicuous spines, subcylindrical or subconical, with tubercular bases; and on the dorsum, in some specimens, may be seen a few others on the median region.

Female of second stage brown, elongated-elliptical, slightly convex. Antennæ six-jointed. Marginal spines as in the adult, but no dorsal spines. Length, about $\frac{1}{25}$ in.

Larva red, flattish, elliptical; marginal spines as in adult, conspicuous. Antennæ of five joints. Abdominal lobes large.

Adult male brown, rather more slender than in *E. hispidus*. Antennæ of ten joints, all hairy; the second joint much thicker than the rest. The last three joints are not so globular as in *E. hispidus*. On the last five joints are several knobbed hairs. Feet slender, hairy; digitules fine hairs. Abdominal spike shortish, broad: at each side of the base a tubercle bearing a pair of longish setæ enclosed in a long filament of white cotton. Eyes four; ocelli two. Length of body, exclusive of spike, about $\frac{1}{25}$ in.

Habitat—On *Atherosperma Novæ-Zelandiæ*, *Melicope ternata*, *Elæocarpus dentatus*, Wellington; *Muhlenbeckia adspersa*, Sumner Road, Lyttelton; Port Hills, Christchurch; and Wellington.

Distinguished from *E. hispidus* by the feather-like segments of the fringe, and by the almost complete absence of dorsal spines on the female. The male is apparently almost identical.

Group III.—HEMICOCCIDINÆ.
N.Z. Trans., Vol. XVI., 1883, p. 128.

Larvæ presenting at the extremity of the abdomen two conspicuous protuberances, or "anal tubercles," as in the following group, *Coccidinæ*; abdominal cleft and lobes absent.

Adult females exhibiting the abdominal cleft and lobes of *Lecanidinæ*. Insects naked or covered with secretion.

The formation of this group has been necessary to include certain insects, *e.g.*, *Kermes*, which are evidently intermediate between the *Lecanidinæ* and the *Coccidinæ*, exhibiting at various stages the characters of each.

SUBDIVISIONS AND GENERA.

Subdivision I.

Adult females naked or covered with horny secretion, without fringe	KERMITIDÆ.
Adult females globular	KERMES.

Subdivision II.

Adult females covered with a test of glassy or waxy secretion	CRYPTOKERMITIDÆ.
Test hard, waxy, with single fringe	POLLINIA.
Test hard, waxy, with double fringe	ASTEROLECANIUM.

None of the insects belonging to this group have as yet been reported as occurring in New Zealand.

As regards the name "Kermes" included in the foregoing list, see N.Z. Trans., Vol. XVII., 1884, p. 17.

Group IV.—COCCIDINÆ.

Adult females variable in form and colour; body segmented, the segments more or less conspicuous; naked, or covered with secretion which is mealy, cottony, felted, or waxy; active or stationary. No abdominal cleft or dorsal lobes; abdomen ending in two more or less conspicuous protruding processes, or "anal tubercles." Mentum, when present, bi- or tri-articulate.

Larvæ exhibiting anal tubercles as in adult.

Adult males of general form of the family; abdominal spike usually short. Antennæ usually of ten joints. Eyes often facetted.

SUBDIVISIONS AND GENERA.

Subdivision I.

Adult females stationary; naked, or covered with cottony or felted secretion; antennæ of not more than seven joints; ano-genital ring with six or eight short hairs; anal tubercles conspicuous. Eyes of male not facetted .... ACANTHOCOCCIDÆ.

Adult female enclosed in closely-felted or waxy sac with double glassy fringe; apodous and without antennæ .... PLANCHONIA.

Adult females enclosed in a felted sac without fringe; retaining feet and antennæ ERIOCOCCUS.

Adult females naked; retaining feet and antennæ .... RHIZOCOCCUS.

Genera not yet represented in New Zealand.

Adult females lying on cushion of cotton; apodous and without antennæ .... NIDULARIA.

Adult females lying on cushion of cotton; retaining feet and antennæ .... GOSSYPARIA.

Adult females enclosed in a cottony sac; apodous; with or without antennæ; excreting from the abdomen a very long cottony appendage .... ANTONINA.

The genus *Acanthococcus*, Signoret (loc. cit., 1874, p. 34), is here united to *Eriococcus*: and the genus *Capulinia*, Signoret (loc. cit., 1874, p. 27), to *Antonina*.

Subdivision II.

Adult females active or stationary; naked, or covered with mealy, cottony, or waxy secretion; antennæ of from six to nine joints; anogenital ring conspicuous, with several long hairs; anal tubercles inconspicuous. Eyes of male sometimes facetted   Dactylopidæ.

Adult females having antennæ of eight joints; anogenital ring with six hairs; naked, or covered with meal or cotton   Dactylopius.

Adult females having antennæ of nine joints; covered with cottony secretion; anogenital ring with six hairs; upper digitules of foot absent   Pseudococcus.

Genera not yet represented in New Zealand.

Adult females having antennæ of six joints; covered with mealy secretion; anogenital ring with six hairs   Ripersia.

Adult females having antennæ of nine joints; anogenital ring with eight hairs; upper digitules of foot present   Puto.

Adult females having antennæ of eight joints; covered with plates of waxy secretion; anogenital ring with six hairs. Eyes of male facetted   Orthezia.

The genera *Westwoodia* and *Boisduvalia*, Signoret (loc. cit., 1874, pp. 337, 338), are here united to *Dactylopius*.

Subdivision III.

Not yet represented in New Zealand.

Adult females active, covered with mealy secretion; antennæ of seven joints; no hairs on anogenital ring. Eyes of male not facetted   Coccidæ.

Adult females having antennæ of seven joints; no hairs on anogenital ring   Coccus.

Subdivision IV.

Adult females active or stationary; naked, or covered with mealy, waxy, or cottony secretion; antennæ of ten or eleven joints; anterior pair of feet similar to the rest; anogenital ring without hairs; anal tubercles inconspicuous. Males with facetted eyes and no ocelli MONOPHLEBIDÆ.

Adult females with elongated antennæ of eleven joints; covered with mealy or cottony secretion; with or without ovisac; rostrum present ICERYA.

Adult females with antennæ of eleven joints; naked, or covered with cottony or mealy secretion; rostrum absent CŒLOSTOMA.

Genera not yet represented in New Zealand.

Adult females having antennæ of eleven joints; naked. Males with several long processes, or tassels, on the abdominal segments MONOPHLEBUS.

Adult females with antennæ of eleven joints; naked. Males without tassels LEACHIA.

Adult females with conical antennæ of eleven joints; covered with cotton ORTONIA.

Adult females with antennæ of ten joints. Males bearing a long mass of silky hairs on the last abdominal segment CALLIPAPPUS.

Adult females with antennæ of ten joints; covered with secretion, partly cottony, partly plates of wax WALKERIANA.

The genera *Drosicha*, Walker (list of Homop. Suppl., 1858, 306, 1) and *Guerinia*, Targioni-Tozzetti (Signoret, loc. cit., 1875, p. 356) are here united to *Monophlebus*.

The genus *Llaveia*, Signoret (loc. cit., 1875, p. 370), is omitted, its affinities being doubtful.

Subdivision V.

Not yet represented in New Zealand.

Adult females with antennæ of variable number of joints; anterior pair of feet abnor-

mally enlarged, deformed; rostrum, mentum, and buccal apparatus absent. Eyes of male facetted .... PORPHYROPHORIDÆ.
Adult females with antennæ of eleven joints; covered with mealy or cottony secretion .... PORPHYROPHORA.
Adult females with antennæ of seven joints; covered with waxy or calcareous secretion .... MARGARODES.

SUBDIVISION I.—ACANTHOCOCCIDÆ.

Female insects exhibiting in all stages the anal tubercles. Young larvæ free, naked, active. Females of second stage active, covered with thin cottony secretion. Adult females stationary; naked, or either resting on or covered with a thick cottony or felted secretion. Anal tubercles in all stages conspicuous, bearing terminal hairs. Anogenital ring inconspicuous, with fine short hairs. Body distinctly segmented.

Male pupæ enclosed in cottony or felted sac. Abdominal spike of adult usually short, with a curved appendage.

Genus: PLANCHONIA, Signoret, Ann. de la Soc. Entom. de France, 1868, p. 282.
Maskell, N.Z. Trans., Vol. XIV., 1881, p. 223.

Adult females enclosed in a sac, or test, of secretion so closely felted as to appear waxy; round the edge of the sac a double fringe of glassy tubes; apodous; antennæ absent. Anal tubercles present in all stages.

The differences between this genus and *Asterolecanium* (see above under *Lecanidinæ*) are not to be made out from external examination of the sac, nor without close investigation. It is possible, indeed, that *Asterolecanium* should be removed from the *Lecanidinæ* and united with *Planchonia*.

53. PLANCHONIA EPACRIDIS, Maskell.
N.Z. Trans., Vol. XIV., 1881, p. 224.
(Plate XII., Fig. 2.)

Test of adult female closely felted, smooth, hard, semitransparent; flat beneath, convex above; elongated-oval, tapering towards the posterior end; completely enclosing the insect except at extreme posterior end, where there is a small orifice;

colour partly green, partly yellow. Round the edge a fringe of long white glassy tubes in double row, one row over the other. Average length, exclusive of fringe, about $\frac{1}{16}$ in.

Test of male unknown.

Adult female filling the test, shrivelling at gestation. Antennæ absent, but represented by circular rings, each bearing four hairs. Feet absent. Anal tubercles small, setiferous. Anogenital ring inconspicuous, with six fine short hairs. Mentum uni-articulate; rostral setæ short. Round the edge of the body a row of figure-of-8 spinneret orifices; on the dorsal surface many circular spinnerets and a number of protruding tubes.

Female of second stage elongated, flattish; colour, reddish-brown; length, about $\frac{1}{20}$ in. Antennæ absent, represented by rings, as in adult. Feet absent. Anal tubercles as in adult. Spinnerets and fringes as in adult.

Young insect elongated-oval, tapering towards anal extremity, flat; anal tubercles clearly distinguishable, setiferous. Antennæ of five joints; the last joint clavate and bearing hairs. Feet normal; upper digitules fine hairs; lower digitules absent. Colour reddish-brown; on the dorsal surface many scattered figure-of-8 spinnerets, from which spring long, curling, white, glassy tubes. Length of insect, about $\frac{1}{40}$ in.

Adult male unknown.

Habitat—On *Leucopogon Fraseri*, Amberley and Sumner, Canterbury; on *Leptospermum scoparium* (manuka) (sparingly), Nelson.

A very pretty little species, resembling in outward appearance *Asterolecanium quercicola*, Signoret, but much smaller, and distinguishable by the presence of the anal tubercles.

Genus: ERIOCOCCUS, Targioni-Tozzetti.

Signoret, loc. cit., 1874, p. 29.

Maskell, N.Z. Trans., Vol. XI., 1879, p. 218.

Adult female enclosed in an elongated sac of white or yellow felted cotton; body elongated, segmented; anal tubercles conspicuous; feet and antennæ present; several rows of conical spines on dorsal surface. Antennæ of six joints.

54. ERIOCOCCUS ARAUCARIÆ, Maskell.
 N.Z. Trans., Vol. XI., 1878, p. 218; Vol. XVI., 1883, p. 134.
 Rhizococcus araucariæ, Comstock; Rept. of Entom., U.S. Agric. Dept., 1881, p. 339.
(Plate XIV., Fig 1.)

Sac of adult female white, cottony, elongated, often aggregated in masses; length, about $\frac{1}{10}$in.

Sac of male similar, but much smaller.

Adult female elongated-oval, convex, segmented; colour yellowish; anal tubercles brown, conspicuous. Length of insect, about $\frac{1}{12}$in. Antennæ of six joints, with some hairs. Feet normal. Anogenital ring inconspicuous, with eight short hairs. On the edge of the body a row of conical spines (spinnerets). After gestation the insect loses its regular oval outline, shrivelling up at one end of the sac.

Young larva and female of second stage similar to adult, but smaller.

Adult male, "a delicate fly-like creature, with two large wings and a pair of long waxen filaments projecting from posterior part of the abdomen; these filaments are very conspicuous, being white, and longer than the body of the insect. Colour of body white, with many irregular markings" (Comstock, loc. cit.).

Habitat in New Zealand—On *Araucaria excelsior* (Norfolk Island pine), Governor's Bay, Canterbury. In America, on same plant.

This insect is not greatly different from *E. buxi*, Signoret; but the sac differs, and there are a few distinguishing characters in the form of the antennæ and feet.

55. ERIOCOCCUS HOHERIÆ, Maskell.
 N.Z. Trans., Vol. XII., 1879, p. 298; Vol. XIX., 1886.
(Plate XIV., Fig. 2.)

Sac of adult female white, cottony, irregularly elliptical, slightly convex, often aggregated in masses; frequently so covered with black fungus as to present the appearance of a minute gall.

Sac of male white, convex, smaller and more elongated than that of the female.

Adult female elongated-oval, convex, red in colour; length, about $\frac{1}{30}$in. Body segmented, tapering rapidly to the posterior extremity; the cephalic segment occupying more than half the length. Anal tubercles conspicuous, apparently two, but on close inspection found to be four, brown in colour; two of them bear longish setæ; all the four are much corrugated and bear many short spiny hairs. Anogenital ring inconspicuous, with eight fine hairs. Antennæ of six short joints, tapering. Foot as if atrophied, the joints small and swollen; digitules all fine hairs. On the body are many small conical spines, which are most numerous on the last two abdominal segments.

Larva free, active, red in colour, flattish, elongated, tapering to the abdominal extremity; length, about $\frac{1}{50}$in. Body segmented; anal tubercles two, conspicuous, setiferous, with some short hairs. Antennæ of six joints; feet normal.

Adult male red, about $\frac{1}{20}$in. long; wings rather narrow, hyaline. Antennæ ten-jointed, hairy. Feet normal. Abdominal spike short, thick, and accompanied by a curved appendage; at each side a tubercle bearing longish setæ.

Habitat—In crevices (and sometimes on surface) of bark of *Hoheria angustifolia*, on hills above the town of Lyttelton. About midsummer individuals may be found which have just completed their sac and have not yet become coated with the black fungus.

The pegtop form and the four anal tubercles of this insect distinguish it from all others of the genus.

56. ERIOCOCCUS MULTISPINUS, Maskell.

Acanthococcus multispinus, Maskell; N.Z. Trans., Vol. XI., 1878, p. 217; Vol. XII., 1879, p. 292; Vol. XVII., 1884, p. 29.

(Plate XV., Fig. 1.)

Sac of adult female yellow, felted, elongated-oval. Length, about $\frac{1}{20}$in. Abdominal extremity open.

Sac of male similar to that of the female.

Adult female pinkish in colour, elongated-oval, convex, filling the sac, shrivelling at gestation; segmented, segments not conspicuous. Length, about $\frac{1}{25}$in. Anal tubercles brownish, conspicuous, setiferous. Anogenital ring small, with eight fine hairs, which are often glued together by cottony secretion, forming a pencil between the tubercles. Antennæ of six joints. Feet having the tibia shorter than the tarsus;* digitules fine hairs. On the dorsal surface are a great number of conical spines, of which the largest are arranged in six longitudinal rows; from the spines sometimes protrude some cottony tubes with an expansion a little below the tip.

Young larva free, active, elongated-oval, flattish; spines as in adult.

Adult male orange-red in colour; length, about $\frac{1}{25}$in. Antennæ of ten joints. Abdominal spike short, thick, with a curved appendage. Feet normal.

Habitat—On *Rubus australis*, Riccarton Bush, Canterbury; on *Knightia excelsa*, *Cyathodes acerosa*, Wellington.

The very numerous conical spines distinguish this species from that known as *Acanthococcus aceris*, Signoret, the European species.

There seems to be no sufficient reason why the genus *Acanthococcus* should have been separated from *Eriococcus*, and they have been here reunited. The only difference mentioned by Signoret is the colour and texture of the sac, an unimportant character in this case.

57. ERIOCOCCUS PALLIDUS, Maskell.
N.Z. Trans., Vol. XVII., 1884, p. 29.
(Plate XV., Fig 2.)

Sac of adult female yellowish-white, elongated-oval, convex, closed at both ends. Length, about $\frac{1}{9}$in.

Sac of male unknown.

Adult female greenish-grey, turning to brown after gestation; filling the sac; shrivelling at gestation. Anal tubercles rather large and conspicuous. Anogenital ring small, with eight (sometimes six?) fine short hairs. Antennæ of six joints. Feet normal, slender; lower digitules narrow and rather long. Body segmented; segments not very distinct. On the middle of each segment a transverse row of small slender conical spines not set closely together. Very many small scattered oval spinneret orifices.

Adult male unknown.

Habitat—On *Myoporum lætum* (ngaio), *Elæocarpus dentatus* (hinau), &c.; throughout the Islands.

---

* As a rule, a tibia shorter than the tarsus characterizes an insect not yet arrived at the adult stage. The genera *Eriococcus* and *Rhizococcus* sometimes present exceptions to this rule. See also, above, the genus *Eriochiton*.

Allied to *E. buxi*, Signoret (European), and *E. multispinus*, ante: but different from both in colour, in the arrangement of the spines and spinnerets, and in the form of the antennae.

Genus: RHIZOCOCCUS, Signoret.

Adult females naked, usually stationary; body segmented; anal tubercles conspicuous. Antennae of six or seven joints. Feet present. Anogenital ring inconspicuous, with fine hairs.

Male pupa enclosed in a cottony sac.

Mr. Comstock proposes (Ann. Rept. of Entom., U.S. Agric. Dept., 1881, p. 339, note) to include in this genus all the species of *Eriococcus*. The organic difference disclosed by the formation of a sac in that genus and the absence of a sac in *Rhizococcus* seems to render the separation of the two necessary.

58. RHIZOCOCCUS CELMISIE, Maskell.

N.Z. Trans., Vol. XVI., 1883, p. 135.

(Plate XVI., Fig 1.)

Adult female deep red in colour, elongated-oval, convex above and flattened below; length about $\frac{1}{17}$ in. The segments of the body are not very distinct. The abdomen ends in two large and conspicuous anal tubercles, each of which bears one strong and fairly long terminal seta and three other spines. The anal ring has eight hairs. Antennae of six joints, sometimes looking like seven. Mentum doubtfully dimerous. The four digitules of the foot are long fine hairs. The tibia is a little shorter than the tarsus.* The trochanter bears one long hair and two short ones. A few large conical spines (spinnerets) are scattered over the body, and a row of smaller ones, like hairs with tubercular bases, runs transversely on each segment; also some circular spinnerets. At the edge of the body, all round, is a row of the large conical spines, which are set in groups of three on the posterior segments, of four or five on the median segments, and almost continuous on the head. When the insect is alive these spines are often agglutinated with cottony secretion so as to give the appearance of a short fringe. The four spiracles are somewhat large and circular.

Adult male unknown.

Habitat—On *Celmisia* sp., Southern Alps, Canterbury.

---

* See note, above, under *Eriococcus multispinus*.

Differs from the European *R. gnidii* in size, colour, and habitat, that species living on the roots of grass, while the New Zealand insect is arboreal. There are also differences in the foot and in the arrangement of the spines and hairs.

It is possible that this insect may, in its latest stage, construct a sac: in that case, it would belong to *Eriococcus*.

59. Rhizococcus fossor, Maskell.
N.Z. Trans., Vol. XVI., 1883, p. 136.
(Plate XVI., Fig. 2.)

Female naked in all stages, but the adult usually buried in a pit.

Male pupa enclosed in a white, elongated, cottony sac, which is about $\frac{1}{20}$in. long.

Adult female greenish-yellow in colour, sometimes brown, stationary; sometimes resting on the leaf, usually partly enclosed in a circular pit; almost circular in outline, flat beneath and slightly convex above; length, about $\frac{1}{15}$in. In the last stage, after gestation, it becomes dark-brown. The cephalic part is smooth; the remainder segmented. The abdomen ends in two very small anal tubercles, which are nevertheless somewhat conspicuous on account of their brown colour. Between them there protrudes a long thick pencil of white cotton. Antennæ short, with six joints, the last joint bearing several long hairs. Feet very small; the femur rather thick; the tibia is shorter than the tarsus by about one-third; the four digitules are long fine hairs. The anal tubercles have not terminal setæ; anal ring inconspicuous. A row of a few conical spines, set far apart, runs round the edge of the body, but none elsewhere, nor any circular spinnerets. There is no sign of a sac in any stage.

Female of the second stage oval, flatter than the adult, and of a rich golden colour; length, about $\frac{1}{40}$in. The segments of the body are somewhat more distinct than in the adult. The anal tubercles are proportionately larger, and bear terminal setæ. Antennæ longer than in the adult, with six joints. Feet also longer. All round the edge runs a row of conical spines, set more closely than in the adult; and from each of these springs a long curly tube of white cotton, making a kind of fringe to the body; each tube is a little dilated at the end, and then tapers to a narrow point. The base of each conical spine is a somewhat large tubercle.

Young larva free, active, elongated, slightly convex, tapering to the anal extremity; colour yellow; length, about $\frac{1}{80}$in. Antennae as in adult, with six joints. Feet somewhat large. Anal tubercles thick, conspicuous, setiferous, with one short hair. On the edge of the body a row of conical spines set far apart, and on the dorsum four other longitudinal rows.

Adult male red in colour, about $\frac{1}{30}$in. long. Antennae of nine joints, all bearing hairs; the last joint nearly globular. Feet slender; digitules fine hairs. Abdominal spike short, thick, with sometimes a curved appendage. A rather strong seta on each side of the base of the spike.

As a rule, the adult female is nearly buried in a circular depression, or pit, formed in the leaf, and with the wall of the pit somewhat curled over it. On the other side of the leaf is a corresponding elevation, of a brown colour. Diameter of pit, about $\frac{1}{18}$in. The abdominal pencil of cotton and the anal tubercles of the female usually protrude at the edge of the pit, probably to attract the male. After gestation, the female disappears within the pit, and the young larvae are also sheltered in it for a while.

Sometimes two females inhabit the same pit.

The females which are not in pits are generally of a dark-red, or brown, colour.

Habitat — On *Santalum Cunninghamii* (maire), Te Aute, Hawke's Bay.

A very distinct species, easily distinguishable by the pits on the leaves.

Subdivision II.—DACTYLOPIDÆ.
The "Mealy-bugs."

Female insects active or stationary; naked, or covered with mealy, cottony, or waxy secretion. Body segmented. Antennæ of from six to nine joints. Feet present. Anogenital ring large, usually conspicuous, with several long hairs. Anal tubercles small, inconspicuous.

Males of general form of the family. Eyes sometimes facetted, usually granular.

Genus: DACTYLOPIUS, Costa.

Adult females having antennæ of eight joints; anogenital ring with six hairs. Naked, or more usually covered with mealy or cottony secretion.

Male pupa enclosed in cottony sac.

60. Dactylopius alpinus, Maskell.

N.Z. Trans., Vol. XVI., 1883, p. 138.

(Plate XVII., Fig. 1.)

Adult female dark-purple in colour, producing a rich red tint in alcohol; body segmented, convex, slightly elongated, stationary; enclosed in a thick mass of white cottony secretion. Length, about $\frac{1}{8}$in. Internal substance very oily. Anal tubercles inconspicuous, thick, with broad bases. Anogenital ring large, with six hairs. Feet normal; upper digitules long fine hairs; lower digitules somewhat broader. Antennæ of eight joints. On the dorsum a number of tubular projecting spinnerets, and others circular. On each of the last three abdominal segments a row of large conical spines.

Female of second stage dark-brown, active, elongated, segmented, flatter than the adult; length, about $\frac{1}{20}$in. Anal tubercles inconspicuous, setiferous. A few conical spines on the posterior segments. Antennæ of six joints. A thin mealy secretion on the body.

Young larva dark-brown, naked, active, elongated, segmented; length, about $\frac{1}{40}$in. Antennæ of six joints. Anal tubercles thick, broad, and more conspicuous than in the adult. A few small spines on the dorsum.

Adult male unknown.

Habitat—On Veronica sp., Upper Waimakariri Valley, Southern Alps.

A species easily distinguished by its cottony sac, its rich colour in alcohol, and its conical spines. It would seem to be intermediate between *Rhizococcus* and *Dactylopius*.

61. Dactylopius calceolariæ, Maskell.

N.Z. Trans., Vol. XI., 1878, p. 218; Vol. XVI., 1883, p. 138.

(Plate XVII., Fig. 2.)

Adult female dull-pink in colour, elongated, distinctly segmented, slightly convex; active; covered with thin mealy secretion; length variable, from $\frac{1}{8}$in. to $\frac{1}{4}$in. Very short cottony appendages sometimes along the edge of the body. Anal tubercles inconspicuous, bearing fine hairs, from which spring two long cottony filaments. Anogenital ring large, with six long hairs which are often glued together by white cotton, forming a pencil between the anal tubercles. Interior substance very oily. Antennæ of eight joints, each bearing hairs. Mentum triarticulate, with a few hairs at the tip. Feet normal.

Female of second stage similar, but smaller. Antennæ of six joints. Anal tubercles somewhat more conspicuous than in the adult.

Adult male unknown.

Habitat—On *Calceolaria*, Christchurch; *Danthonia*, Stewart Island; *Phormium tenax*, Christchurch.

The large size, and the absence of long cottony marginal appendages, distinguish this species from the European *D. adonidum*.

62. Dactylopius glaucus, Maskell.

N.Z. Trans., Vol. XI., 1878, p. 219; Vol. XVII., 1884, p. 30.

(Plate XVII., Fig. 3.)

Adult female green, sometimes brownish-red, elongated, distinctly segmented, slightly convex; active; covered with thin mealy secretion. Length, averaging $\frac{1}{14}$in. Body oval, tapering somewhat to the posterior extremity. Anal tubercles inconspicuous, each bearing fine hairs and a long cottony filament. Anogenital ring large, with six hairs often forming with cotton a protruding pencil. A few cottony appendages sometimes round the edge of the body, often absent. Antennæ of eight joints, each bearing hairs. Feet normal; lower digitules rather broad.

Sac of male pupa narrow, cylindrical, white, cottony, open at the posterior end. Length, about $\frac{1}{8}$in.

Adult male about $\frac{1}{20}$in. long; brown, covered when newly hatched with white meal. Body rather thick; abdominal spike short. Antennæ of ten joints, hairy; the last eight joints equal to each other. Feet slender, hairy; upper digitules long, fine; lower digitules short.

Habitat—On *Panax, Rubus, Coprosma, Pittosporum, Piper excelsum*, &c.; throughout the Islands: also frequently on fruit-trees.

A species more nearly resembling the ordinary "mealy bug," *D. adonidum*, than any other in New Zealand; but differing in colour, in the absence of long marginal appendages, and in the form of the foot and antennæ.

63. DACTYLOPIUS POÆ, Maskell.

N.Z. Trans., Vol. XI., 1878, p. 220.

(Plate XVIII., Fig. 1.)

Adult female pink, covered with thin white meal; slightly elongated, sometimes globular; flat beneath, convex above; segmented, the segments indistinct. Length reaching about $\frac{1}{10}$in. Antennæ of eight joints, very short. Feet normal, very short; upper digitules short, lower digitules absent (?). Anal tubercles extremely small and inconspicuous; each has three conical spines, but no hairs. Anogenital ring large, with six hairs. On the dorsum are numbers of small circular spinnerets.

Adult male unknown.

Habitat—On the common tussock grass, *Poa anceps* (*australis?*), Mount Grey Downs and Port Hills, Canterbury; either just above the ground, or more often an inch or two below the surface.

A species clearly distinct in form and habit.

Genus: PSEUDOCOCCUS, Westwood.

Adult females covered with cottony secretion; stationary; antennæ of nine joints; anogenital ring conspicuous, with six hairs; upper digitules of the foot absent.

64. Pseudococcus astelle, Maskell.
N.Z. Trans., Vol. XVI., 1883, p. 139.
(Plate XVIII. Fig. 2.)

Adult female about $\frac{1}{10}$ in. long, yellowish-brown, covered with a not very abundant white cotton; segmented; anal tubercles inconspicuous; anal ring with six hairs. Antennae with nine joints, of which the third, fourth, and fifth are the longest; the second, sixth, and ninth a little shorter; the first, seventh, and eighth the shortest. The fourth, fifth, and sixth are the narrowest, the two ends of the antennae being thicker than the middle. The eighth joint is a little expanded at the tip; and the ninth is fusiform, with a shallow depression at the extremity. All the joints have a few long hairs, and on the eighth is one a good deal stronger than the others. The legs have the tibiae twice as long as the tarsus; the claw is slender, and has no tooth on the inner edge. There are only two digitules (the lower pair), which are long and fine. The trochanter bears one short bristle. The whole leg is slender and long. The eyes are tubercular and smooth, showing after maceration in potash a small dark terminal spot. The body is covered with a number of spinnerets of two kinds: those with simple concentric circles are the largest, and are found all over the integument; the others are multilocular, and are placed in groups at the edges of the segments and also in great numbers at the cephalic and abdominal extremities. Interspersed with these spinnerets are several hairs, mostly very short, but on the head are some pretty long. From the anal tubercles spring two strong setae with tubercular bases, not very long. The mentum is dimerous, and bears a few hairs on the tip. In the groups of spinnerets at the edges of the segments are found a few small conical spines. The four spiracles are small and simple.

Adult male unknown.

Habitat—On *Astelia* sp., in forests, Hawke's Bay.

Allied to *P. Mespili*, Geoffroy; but differs in the antennae, feet, and spinnerets.

Subdivision III.—COCCIDÆ.

Adult females active, covered with mealy secretion; antennæ of seven joints; no hairs on anogenital ring. Eyes of male not facetted.

This subdivision, which includes the single genus *Coccus*, of which there would seem to be not more than one distinct species —*Coccus cacti* (the cochineal insect)—and two or three varieties, has not yet any representatives in New Zealand.

Subdivision IV.—MONOPHLEBIDÆ.

Adult females active or stationary; naked, or covered with mealy, cottony, or waxy secretion; segmented; antennæ of ten or eleven joints; anterior pair of feet similar to the rest; anogenital ring without hairs; anal tubercles inconspicuous.

Males with facetted eyes and no ocelli.

Strictly speaking, the wings of the males of this group should, according to its name, present only a single nervure. This, however, is not the case, or, rather, it should be said that the nervure is precisely similar to that of all other Coccids, branching once, so that it cannot form a distinctive character. Possibly the name of *Monophlebus* was originally given by Leach to an abnormal or imperfect specimen.

Genus : ICERYA, Signoret.

Adult females having antennæ of eleven joints; covered with thin mealy secretion or with cotton; stationary; with or without ovisac. Rostrum and mentum present. Segmentation inconspicuous.

Adult males without tassels on the abdomen; antennæ with two dilations on each joint.

Two species only of this genus are at present known, the one described below and another, *I. sacchari*, infesting sugar-canes in Mauritius. The male of the latter is unknown. Possibly researches in Australia might result in the discovery of others.

65. Icerya Purchasi, Maskell.
   N. Z. Trans., Vol. XI., 1878, p. 221; Vol. XVI., 1883, p. 140; Vol. XVII., 1884, p. 30; Vol. XIX., 1886, p. 45.

The " Cottony-cushion Scale."
(Plate XIX.)

Adult female dark reddish-brown, covered with a thin powdery secretion of yellowish meal, and with slender glassy filaments; stationary at gestation, and gradually raising itself on its head, lifting the posterior extremity until nearly perpendicular, filling the space beneath it with thick white cotton, which gradually extends for some distance behind it in an elongated, white ovisac, longitudinally corrugated; ovisac often much longer than the insect, and becoming filled with oval red eggs. Length of female, about $\frac{1}{4}$in., reaching sometimes nearly $\frac{1}{3}$in. Body previous to gestation lying flat on the plant, the edge slightly turned up; on the dorsum a longitudinal raised ridge, forming one or more prominences. Insect covered all over with numerous minute fine hairs, most thickly on the thoracic region; round the edge these hairs are longer, and are arranged in tufts somewhat closely set; the tufts are black, and contain from twenty to thirty hairs in each. Amongst the hairs in the tufts are several protruding tubular spinnerets, having on the outer end a kind of multiglobular ring or crown; from these proceed cylindrical, glassy, straight tubes as long as the tufts of hair. Long, fine, glassy, delicate filaments, as long as the body of the insect, radiate from the edge all round; but these, being very fragile, are often irregular, or absen During gestation thick, short, cottony processes form at the edge of the thorax, seemingly attached to the feet. Antennæ of eleven joints, very slightly tapering; each joint bearing hairs. Feet normal, somewhat thick. Rostrum not long; mentum triarticulate. Procreation commencing soon after the first formation of the ovisac, the eggs being ejected into the sac as it grows; ovisac at completion containing sometimes as many as 350 eggs; ovisac convex above, sometimes irregularly split, more often nearly conical, divided by several regular longitudinal grooves or ribs.

Female of second stage dark-red, elongated, slightly convex, active, covered with thin meal, or short curly cotton. Body hairy with marginal tufts and spinnerets, as in adult. Anal tubercles inconspicuous, but the abdomen exhibits three small lobes on each side, from which spring six short setæ. Antennæ of nine nearly equal joints, hairy. Feet normal, thick. Several radiating, fine cottony filaments. Length of insect variable, from $\frac{1}{10}$in. to $\frac{1}{6}$in. The dorsum exhibits the longitudinal raised ridge, but less conspicuously than in the adult.

Young larva, about $\frac{1}{24}$in. long, dark-red, elongated, flattish, active; covered with yellow cottony down. Antennæ of six joints, hairy; the last joint is much the largest, clavate, apparently four-ringed, bearing four long hairs. Feet slender; digitules short, fine hairs. Eyes prominent, tubercular. Mentum biarticulate. Anal tubercles represented by three small processes at each side of the abdominal extremity, each process bearing a very long seta. Six longitudinal rows of circular multilocular spinnerets, four on the dorsum and one on each edge. Alternating with these are rows of hairs with tubercular bases.

Adult male large, the length slightly varying; some specimens reach ½in.; expanse of wings, ⅓in.; length of antennæ, about ¼in. Body red, with a shining, diamond-shaped, black patch on the dorsal surface of the thorax; legs and antennæ black. Wings dark-brown with (in some lights) a bluish tinge, marked with oblique, narrow, wavy stripes; main nervure red, branching once; there are also two longitudinal, whitish, narrow bands.* Antennæ very long and slender, with ten joints, which may easily be taken for nineteen, for, after the first, which is short, round, and simple, all the other nine have two dilated portions with a constriction in the middle, and on each dilation is a ring of very long hairs, giving the antenna a feathery appearance.† Eyes very large and prominent, almost pedunculated, brown, divided into numerous semiglobular facets. Feet long and very hairy; coxæ short and thick, tibiæ long and slender, claw thin; upper digitules absent, lower pair only short bristles. Abdomen slender, segments somewhat distinct; on each segment some hairs; the last segment ends in two thick, conspicuous, cylindrical processes, which, in side view, are seen to incline upwards, and beneath them is the short, conical spike sheathing the penis. Penis red, longish, tubular, and thick, with many recurved short hairs, and at the end a ring of short spines. Each of the two processes on the last segment bears three or four long setæ, but there do not appear to be any of the long cottony appendages seen in the males of most Coccids.

Habitat — On wattle, pine, orange, lemon, cypress, rose, gorse, grass, and, in fact, on almost every kind of native and introduced plants, Nelson, Hawke's Bay, Auckland. It will probably appear also elsewhere, but the climate of Canterbury and Otago may prove too cold in winter for it.

Allied to *I. sacchari*, Guérin, which damages sugar-canes in Mauritius; but differing in the formation of the ovisac, the

---

* Signoret (Ann. de la Soc. Ent. de France, 1875), under the genus *Monophlebus*, speaks of "les plis hyalins" as existing also in the wings of the males of that genus.

† Misled by similar appearances, Burmeister and Westwood assign twenty-five joints to the male antenna of *Leachia fuscipennis*.

presence of the marginal tufts and spinneret tubes in the female, and in other particulars. The male of *I. sacchari* has not been described. The male of *I. Purchasi* is probably quite distinct.

This species is supposed to have come originally from Australia. It has been very injurious to orange and lemon trees at the Cape of Good Hope and in California. In Auckland it has destroyed whole orchards of the same trees, and in Nelson and Hawke's Bay it is a dreadful pest on all kinds of plants.

Tree-growers should especially beware of this insect, and the best plan to adopt would be to burn at once any tree found infested with it.

Genus: CŒLOSTOMA, Maskell.

N.Z. Trans., Vol. XII., 1889, p. 294.

Adult females with antennæ of eleven joints; segmented; naked, active; at gestation becoming stationary and enclosed in a thick mass of white cottony secretion. Anal tubercles absent or inconspicuous. Rostrum and mentum absent in the adult female.

Males with prominent, facetted eyes; ocelli absent. Abdomen without tassels.

In one New Zealand species the female in the second stage is stationary, enclosed in a thick, hard mass of waxy secretion, of which some account is given in Chap. III. The other species included in this work inhabits a remote and scarcely visited district, and the female has not hitherto been found; it is not possible therefore to include the excretion of wax and the stationary position amongst the generic characters at present.

This genus is allied to *Porphyrophora*, Brandt, and to *Monophlebus*, Leach; differing from the former by the presence of an œsophagal orifice, and from the latter by the absence of tassels on the abdomen of the male. In the genus *Ortonia*, Signoret, the female possesses a rostrum and mentum.

66. CŒLOSTOMA ZEALANDICUM, Maskell.

N.Z. Trans., Vol. XII., 1879, p. 294; Vol. XIV., 1881, p. 226; Vol. XVI., 1883, p. 141.

(Plate XX.; Plate XXI., Fig. 1.)

Adult female brick-red, elongated, distinctly segmented, convex; length about ½in.; before gestation active, naked or covered with thin meal; during gestation stationary, enclosed in a thick mass of white cotton. Antennæ of eleven joints, tapering to the tip; the joints all nearly equal, and bearing several hairs. Feet black, short, strong; all the pairs placed somewhat forward; digitules absent, a short seta at the base of the claw; on the inner edge of the tibia and tarsus a fringe of strong hairs; on the trochanter a long hair. Rostrum and mentum absent; mouth represented by a minute orifice between the second pair of feet. Eyes very minute, tubercular, projecting, placed behind the antennæ. Anal tubercles absent; anogenital opening small, elliptical, simple, hairless. Body flattish beneath, more convex above; segments distinct, generally smooth, sometimes bearing hard projecting callosities. Skin covered with minute hairs interspersed with circular spinnerets.

Female of second stage deep-red in colour, nearly globular, very indistinctly segmented; stationary, enclosed in a thick, hard solid test of yellow wax; tests attaining sometimes the size of a large pea; enclosed insect averaging $\frac{1}{3}$in. in length. Insect filling the test; skin smooth, very thin; general appearance like a hard, round, smooth ball. Anal tubercles absent; anogenital ring small, simple, hairless : from this ring springs often a tuft of short white cotton, and a long white cottony seta protruding through an orifice in the test. Rostrum and mentum present, very small; mentum triarticulate, with a few hairs at the tip. Antennæ very short, of eight joints, conical, tapering to the tip, hairy. Feet atrophied, bloated-looking, apparently only consisting of a very short femur, tibia, and tarsus; digitules two, very small. Skin covered with a great number of circular spinnerets of two sizes, the larger ones simple, the smaller multilocular. Tracheæ very large; spiracular orifices containing brown tubes with beaded extremities on the inner end. Anal extremity dark-brown, the anal ring in the centre; spinnerets here very numerous, converging towards the anus. Insect in this stage emitting a strong, persistent, and fetid odour.

Young larva brick-red, elongated, active, naked; length, about $\frac{1}{24}$in. Antennæ of six joints, hairy; last joint the largest, clavate. Eyes and feet as in adult, but there is no fringe on the tibia and tarsus. Rostrum and mentum large, conspicuous. Skin covered with circular spinnerets and minute hairs; the spinnerets are most numerous on the abdomen. Anogenital ring apparently folded. At the extremity of the abdomen two long hairs springing from quadrate tubercular bases which represent the anal tubercles.

Adult male red or purplish in colour, wings bluish-purple with red nervures; length, about $\frac{1}{3}$in.; width across expanded wings, about $\frac{1}{2}$in. Eyes large, prominent, facetted. Antennæ of ten joints, each joint bearing many hairs. Feet long, slender, with somewhat large trochanter; on the tibia and tarsus a fringe of hairs as in the female. Upper digitules two long fine hairs; lower digitules absent. The nervure of the wings branches twice at least. Haltere large, inflated, sac-like; bearing four curling setæ. Abdomen distinctly segmented, each segment bearing many fine short hairs and several small circular marks. Abdominal spike short, broad, bivalvular; penis protruding as a long soft white tube covered with minute recurved hairs.

Habitat—For the adult female and larva the trunks of trees and shrubs in forests, or rocks and bushes in open country, Otago, Nelson, Canterbury, Wellington. For the second stage the stems and roots of *Muhlenbeckia adpressa (complexa ?)*; Sumner Road, Lyttelton; Evans Bay, Wellington: *Rhipogonum scandens* (supplejack); Riccarton Bush, Canterbury; Nelson; Wellington. On *Muhlenbeckia* the waxy tests are often largest and most numerous underground. Male insects sometimes found clustering, attached to females.

This is a very large and peculiar species, its transformations and changes of secretion being abnormal. It cannot be said to be greatly harmful; but the odour of the second stage is unpleasant.

67. CŒLOSTOMA WAIROENSE, Maskell.

N.Z. Trans., Vol. XVI., 1883, p. 141.

(Plate XXI., Fig. 2.)

Adult female, female of second stage, and larva unknown.

Adult male very nearly resembling that of *C. zealandicum*; body red or purplish, wings blue with red nervures. Length of body, about $\frac{1}{6}$in. Eyes prominent, facetted. Antennæ of ten joints, slender, with fine hairs. Feet as in *C. zealandicum*, but with fewer hairs. Digitules twenty-four, all springing from the claw, none from the tarsus. Haltere, abdominal spike, penis, and abdomen as in *C. zealandicum;* but the circular marks on the segments in this species are multilocular.

Male pupa bright-red, enclosed in a cylindrical sac of white cotton. Pupæ occurring in numerous colonies.

Habitat — On *Phormium tenax, Leptospermum scoparium* (manuka), Northern Wairoa, Auckland.

The female of this species will, when found, probably nearly resemble that of *C. zealandicum:* at present the great number of digitules on the foot of the male sufficiently distinguish it.

INDEX OF PLANTS AND THE COCCIDIDÆ ATTACKING THEM.

This list is of course only approximate, as insects may at any time be found on other trees than those given.

Icerya Purchasi is omnivorous, and it has not been thought necessary to repeat it here against every plant in the list; but it has been set against some.

The adult female of *Cœlostoma zœlandicum* may also be found wandering over numbers of native trees in forests.

In hothouses and greenhouses all sorts of plants are liable to attack.

Plants.	Insects.
Alsophila Colensoi (fern)	Lecanium mori.*
Apple	Mytilaspis pomorum.‡
"	Dactylopius glaucus.*‡
Apricot	Mytilaspis pomorum.‡
Ash	" pomorum.‡
Asplenium bulbiferum (fern)	Chionaspis dubia.*†
" lucidum	" dubia.*†
Astelia Cunninghamii	Mytilaspis cordylinidis.*
" "	" epiphytidis.
" "	Fiorinia asteliæ.*†
" "	" stricta.
" "	Pseudococcus asteliæ.
Atherosperma Novæ-Zælandiæ	Aspidiotus atherospermæ.
" "	Mytilaspis pyriformis.*
" "	Fiorinia asteliæ.*
" "	Ctenochiton viridis.*†
" "	Inglisia patella.
" "	Eriochiton spinosus.
Bavardia	Lecanium maculatum.*!‡
Box	" hesperidum.*†‡
Brachyglottis repanda	Fiorinia minima.
" "	Ctenochiton flavus.
" "	" fuscus.*†
Budlæia salicina	Aspidiotus budlæiæ.
Calceolaria	Dactylopius calceolariæ.*†‡
Camellia	Aspidiotus camelliæ.
"	Lecanium hemisphæricum.*†‡
"	" hesperidum.*!‡
"	" oleæ.*†‡
"	Pulvinaria camellicola.*†‡
Carpodetus serratus	Aspidiotus carpodeti.
Cassinia leptophylla (tauhine)	Lecanium oleæ.*†‡
Celmisia	Rhizococcus celmisiæ.*
Coprosma	Aspidiotus nerii.
"	Mytilaspis pyriformis.*
"	Chionaspis dubia.*
"	Fiorinia asteliæ.*
"	Ctenochiton perforatus.*†

* Unsightly: spoiling appearance of plant.
† Usually accompanied by much black fungus.
‡ Likely to do much injury to plant.

Plants.	Insects.
oprosma	Ctenochiton viridis.* †
„	Inglisia patella.
„	Dactylopius glaucus.*
Cordyline australis (cabbage-tree)	Mytilaspis cordylinidis.* ‡
„	Fiorinia stricta.* ‡
„ indivisa (large-leaved cabbage-tree)	Mytilaspis cordylinidis.* ‡
„ „	Fiorinia stricta.* ‡
Corynocarpus lævigata (karaka)	Aspidiotus nerii.
Cotoneaster microphylla	Mytilaspis pomorum.‡
Cyathea Smithii (tree-fern)	Ctenochiton depressus.*
Cyathodes acerosa	Poliaspis media.*
„	Eriococcus multispinus.*
Cypress	Icerya Purchasi.* ‡
Danthonia (grass)	Dactylopius calceolariæ.*
Dendrobium	Fiorinia stricta.*
„	Ctenochiton elongatus.* †
Drimys colorata	Mytilaspis drimydis.
„ „	Inglisia patella.
Dysoxylon spectabile	Aspidiotus dysoxyli.
„ „	Mytilaspis pyriformis.*
„ „	Chionaspis dysoxyli.* ‡
Earina	Ctenochiton elongatus.
„	Fiorinia stricta.*
Elæocarpus dentatus (hinau)	Ctenochiton elæocarpi.
„ „ „	„ flavus.
„ „ „	Inglisia ornata.*
„ „ „	Eriococcus pallidus.
Eucalyptus	Mytilaspis cordylinidis.*
Euonymus	Aspidiotus camelliæ.* ‡
Ferns, various	Chionaspis dubia.
„ „	Poliaspis media.
„ „	Ctenochiton depressus.*
„ „	Lecanium mori.
„ „	Dactylopius glaucus.* ‡
Gahnia	Mytilaspis cordylinidis.*
Geniostoma ligustrifolia	Ctenochiton elongatus.*
Gooseberry	Fiorinia grossulariæ.
Gorse	Icerya Purchasi.* ‡
Grasses, various	Dactylopius pom.
„ „	Icerya Purchasi.* ‡
Hawthorn	Mytilaspis pomorum.‡
Hedycarya	Fiorinia stricta.
Hoheria angustifolia	Eriococcus hoheriæ.
Holly	Lecanium hesperidum.* † ‡
Hymenanthera crassifolia	Ctenochiton hymenantheræ.* †
Ivy	Lecanium hesperidum.* † ‡
Knightia excelsa	Eriococcus multispinus.*
Laurel	Lecanium hesperidum.* † ‡
Lemon	Aspidiotus coccineus.* ‡
„	Icerya Purchasi.* ‡
Leptospermum scoparium (manuka)	Mytilaspis leptospermi.
„ „ „	Ctenochiton flavus.
„ „ „	Inglisia leptospermi.
„ „ „	„ ornata.
„ „ „	Planchonia epacridis.
„ „ „	Cœlostoma wairoense.*
Leucopogon Fraseri	Poliaspis media. *
„ „	Planchonia epacridis.

---

* Unsightly : spoiling appearance of plant.
† Usually accompanied by much black fungus.
‡ Likely to do much injury to plant.

SCALE-INSECTS.

Plants.	Insects.
Lilac	Mytilaspis pomorum.‡
Melicope ternata	Eriochiton spinosus.
Metrosideros tomentosa (pohutukawa)	Lecanochiton metrosideri.* †
" robusta (rata)	Mytilaspis metrosideri.
	Lecanochiton metrosideri.* †
Muhlenbeckia adspersa	Fiorinia stricta.
" "	Eriochiton spinosus.
" "	Cœlostoma zælandicum.
Myoporum lætum (ngaio)	Eriococcus pallidus.
Myrtle	Lecanium hesperidum.* † ‡
Norfolk Island pine (Araucaria)	Eriococcus araucariæ.* † ‡
Olearia Haastii	Eriochiton hispidus.* †
Orange	Aspidiotus coccineus.* ‡
"	Chionaspis citri.
"	Lecanium hesperidum.* † ‡
"	" oleæ.* † ‡
"	Icerya Purchasi.* ‡
Orchids (hothouse)	Aspidiotus epidendri.* ‡
" "	" nerii.* ‡
" "	Dactylopius glaucus.* ‡
Palms (hothouse)	Aspidiotus epidendri.* ‡
" "	" nerii.* ‡
" "	Dactylopius glaucus.* ‡
Panax arboreum	Fiorinia minima.
" "	Ctenochiton flavus.* †
" "	" fuscus.* †
" "	" perforatus.* †
" "	" viridis.* †
" "	Dactylopius glaucus.*
Parsonsia	Chionaspis minor.
Peach	Mytilaspis pomorum.‡
Pear	Diaspis santali.
"	Mytilaspis pomorum.‡
Pellæa rotundifolia (fern)	Chionaspis dubia.
Phormium tenax (New Zealand flax)	Mytilaspis cordylinidis.*
" " "	Fiorinia stricta.*
" " "	Dactylopius calceolariæ.*
" " "	Cœlostoma wairoense.*
Phymatodes Billardieri (fern)	Mytilaspis phymatodidis.*
Pines and firs	Icerya Purchasi.* ‡
Piper excelsum	Ctenochiton piperis.
" "	Dactylopius glaucus.*
Pittosporum (various)	Fiorinia asteliæ.
" "	Ctenochiton perforatus.* †
" "	" viridis.* †
" "	Dactylopius glaucus.* †
Plagianthus	Ctenochiton depressus.*
Plum	Diaspis santali.‡
"	Mytilaspis pomorum.‡
Poa anceps (tussock-grass)	Dactylopius poæ.
Rhipogonum scandens (supplejack)	Chionaspis minor.*
" "	Cœlostoma zælandicum.
Rose	Diaspis rosæ.* ‡
"	Icerya Purchasi.* ‡
Rubus australis (bush-lawyer)	Chionaspis dubia.*
" "	Ctenochiton perforatus.* †
" "	" viridis.* †
" "	Eriococcus multispinus.

* Unsightly; spoiling appearance of plant.
† Usually accompanied by much black fungus.
‡ Likely to do much injury to plant.

Plants.	Insects.
Rubus australis (bush-lawyer)	Dactylopius glaucus.*
Santalum Cunninghamii (maire)	Diaspis santali.*
" "	Rhizococcus fossor.
Sophora tetraptera (kowhai)	Aspidiotus sophoræ.
Sweetbriar	Icerya Purchasi.* ‡
Sycamore	Mytilaspis pomorum.
Thorn	„ pomorum.‡
Various greenhouse or hothouse plants	Aspidiotus epidendri.* ‡
„ „ „ „	„ nerii.* ‡
„ „ „ „	Diaspis Boisduvalii.* ‡
„ „ „ „	Lecanium hemisphæricum.* † ‡
„ „ „ „	„ hesperidum.* † ‡
„ „ „ „	„ hibernaculorum.* † ‡
„ „ „ „	„ mori.
„ „ „ „	„ oleæ.* † ‡
„ „ „ „	Dactylopius glaucus.* ‡
„ „ „ „	Icerya Purchasi.* ‡
Veronica (various)	Poliaspis media.
„ „	Lecanium hesperidum.* †
„ „	Dactylopius alpinus.*
Vitex littoralis (puriri)	Aspidiotus carpodeti.
Wattle (various)	„ epidendri.*
„ „	Diaspis Boisduvalii.*
„ „	Icerya Purchasi.* †
Weeping-willow	Aspidiotus camelliæ.

* Unsightly: spoiling appearance of plant.
† Usually accompanied by much black fungus.
‡ Likely to do much injury to plant.

INDEX OF GROUPS, SUBDIVISIONS, GENERA, AND SPECIES INCLUDED IN THIS WORK.

Names in italics are synonyms.   *Signifies that the insect has not yet been found in New Zealand.

	Page
ACANTHOCOCCIDÆ	88, 91
Acanthococcus	95
multispinus	94
Aclerda*	63
Antonina*	88
Aonidia*	39
Aspidiotus	39, 40
atherospermæ	40
aurantii	42
Bouchei	44
budlæiæ	40
camelliæ	41
carpodeti	42
citri	42
coccineus (the ran e-scale)	42
conchiformis	51
dysoxyli	43
epidendri	44
falciformis	51
juglandis	51
nerii	44
pomorum	51
pyrus-malus	51
rosæ	47
sophoræ	45
Asterolecanium*	64, 87, 91
quercicola*	92
Boisduvalia	89
Callipappus*	90
Calymnatus hesperidum	80
Calypticus hesperidum	80
Capulinia*	89
Carteria*	62
Ceroplastes*	62
Chermes epidendri	44
filicum	80
hibernaculorum	81
oleæ	82
Chionaspis	39, 54
citri	54
dubia	54
dysoxyli	55
euonymi	54
minor	56
COCCIDÆ	89, 103
COCCIDINÆ	38, 88
Coccus* (the cochineal-insect)	89
hesperidum	80
Cœlostoma	90, 107
wairoense	109
zœlandicum	107

	Page
CRYPTO-KERMITIDÆ	87
Ctenochiton	62, 65
depressus	66
elæocarpi	67
elongatus	68
flavus	68
fuscus	70
hymenantheræ	71
perforatus	72
piperis	73
spinosus	86
viridis	74
DACTYLOPIDÆ (the "mealy-bugs")	89
Dactylopius	89
alpinus	99
calceolariæ	100
glaucus	100
poæ	101
DIASPIDINÆ	37, 39
Diaspis	39, 45
Boisduvalii	46
Bouchei	44
gigas	58
rosæ	47
santali	47
Drosicha*	90
Ericerus*	62
Eriochiton	63, 84
hispidus	84
spinosus	86
Eriococcus	88, 92
araucariæ	93
hoheriæ	93
multispinus	94
pallidus	95
Eriopeltis*	63
Fairmairia*	62
Fiorinia	39, 57
asteliæ	58
grossulariæ	59
minima	59
stricta	60
Gossyparia*	88
Guerinia	90
HEMICOCCIDINÆ	38, 87
Icerya	90, 104
Purchasi (the cottony-cushion scale)	104
sacchari*	106

	Page		Page
Inglisia	62, 75	Mytilaspis—*continued.*	
leptospermi	75	metrosideri	50
ornata	76	phymatodidis	51
patella	78	*pomicorticis*	52
Kermes*	87	pomorum (the apple-scale)	51, 52
camelliæ	41	pyriformis	53
KERMITIDÆ*	87	Nidularia*	88
Leachia*	90	Orthezia*	89
LECANIDÆ	63, 79	Ortonia*	90
LECANIDINÆ	38, 62	Parlatoria*	39
Lecanium	63, 79	Philippia*	63
depressum	79	*Physokermes* *	63
hemisphæricum	80	Planchonia	64, 88, 91
hesperidum (the holly and ivy scale)	80	epacridis	91
		Poliaspis	39, 56
hibernaculorum	81	media	57
maculatum	81	Pollinia*	61, 87
mori	82	Porphyrophora*	91
oleæ (the "black scal")	82	PORPHYROPHORIDÆ*	91
Lecanochiton	62, 64	Pseudococcus	89, 101
metrosideri	64	astoliæ	102
LECANOCOCCIDÆ	63, 84	Pulvinaria	63, 83
LECANODIASPIDÆ	62, 63	camellicola	83
Lecanodiaspis*	63, 64	Puto*	89
Lecanopsis*	63	Rhizococcus	88, 96
Leucaspis*	39	*araucariæ*	93
Lichtensia*	63	celmisiæ	96
Llaveia *	90	fossor	97
Margarodes*	91	Ripersia*	89
MONOPHLEBIDÆ	90, 104	Signorctia*	63
Monophlebus*	90	Targionia*	39
Mytilaspis	39, 48	*Uhleria*	57
cordylinidis	48	*gigas*	58
drimydis	49	Vinsonia*	62
epiphytidis	49	Walkeriana*	90
leptospermi	50	*Westwoodia* *	89

DESCRIPTIONS OF PLATES.

Plate I.

Fig.
1. Eggs of Coccididæ.

2. Larva of Diaspidinæ.

3. Diagram of life-history of female Diaspidinæ. *a*, pellicle of larva; *b*, second pellicle; *c*, adult female somewhat shrivelled after egg-laying; *d*, eggs; *e*, secreted matter forming the "scale." In this diagram the scale is shown as overturned.

4. Types of various spinnerets. *a*, simple orifices (*Mytilaspis*); *b*, double orifice (*Planchonia*); *c*, multilocular orifice (*Ctenochiton*); *d*, multilocular orifice (*Cœlostoma*); *e*, simple protruding spinneret (*Aspidiotus*); *f*, serrated protruding spinneret (*Mytilaspis drimydis*); *g*, protruding spinneret (*Acanthococcus*); *k*, conical spiny spinneret and cottony secretion (*Rhizococcus*); *m*, lanceolate spinneret and glassy secretion (*Inglisia*); *n*, group of abdominal spinnerets, with detached orifices (*Mytilaspis*); *p*, double multilocular orifice (*Cœlostoma*); *s*, coronetted spinnerets and part of glassy secretion (*Icerya*).

5. Rostra. *a*, simple rostrum of Diaspidinæ; *b*, rostrum and trimerous mentum of Coccidinæ.

6. Foot of a female insect. *c*, coxa; *tr*, trochanter; , femur; *ti*, tibia; *ta*, tarsus; *cl*, claw.

7. Foot of a male insect. The letters as in Fig. 6.

8. Foot with digitules, upper and lower.

9. Type of female antenna (*Ctenochiton*).

10. Type of female antenna (*Dactylopius*).

11. Type of antenna of larva (*Icerya*).

12. Type of male antenna (*Florinia*).

13. One joint of antenna of Monophlebidæ (*Icerya*).

14. Diagram of head of male of Lecanidinæ (after Signoret), both dorsal and ventral aspects being shown together. *a*, antennæ; *e, e*, true eyes, the lower pair being on the ventral surface in the place of the mouth; *oc, oc*, ocelli.

15. Head of male Icerya, with facetted eyes.

16. Wing of male insect. *n*, nervure.

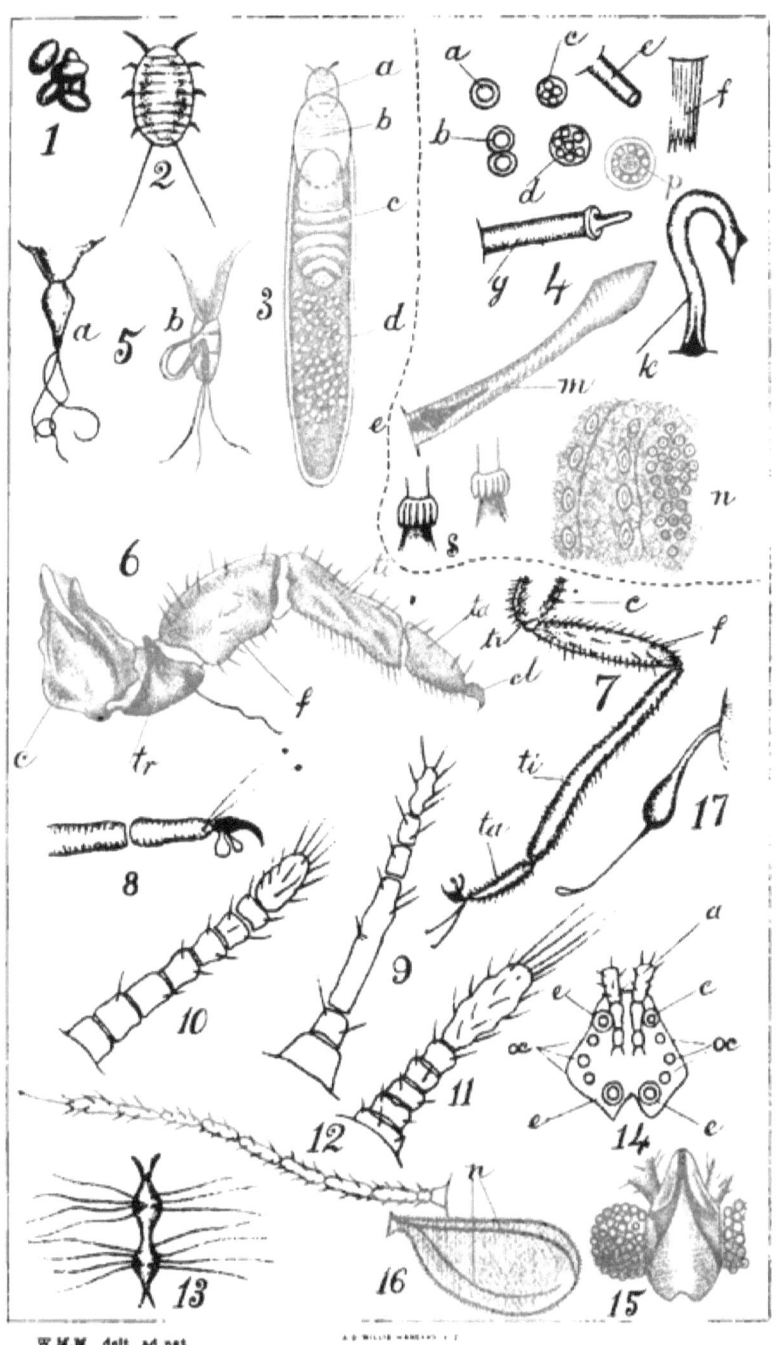

PLATE I.

PLATE II.

Fig.
1. Types of anogenital rings. *a*, Diaspidinæ; *b*, Lecanidinæ; *c*, Acanthococcidæ; *d*, Dactylopidæ; *e*, Monophlebidæ.

2. Types of last abdominal segments of female. *a*, Diaspidinæ; *b*, Lecanidinæ; *c*, Acanthococcidæ; *d*, Dactylopidæ; *e*, Monophlebidæ.

3. Types of last abdominal segments of male, and sheath of the penis. *a*, Diaspidinæ; *b*, Lecanidinæ (*Ctenochiton*); *c*, Acanthococcidæ; *d*, Lecanidinæ (*Inglisia*); *e*, Dactylopidæ; *f*, Monophlebidæ (*Icerya*).

4. Respiratory system. *a*, spiracle of *Lecanium* (after Targioni); *b*, spiracle of *Coccus* (after Targioni); *c*, diagram of arrangement of the four spiracles and the tracheal tubes (Lecanidinæ) (s. spiracles); *d*, spiracle and trachea of Cœlostoma.

5. Types of spines and hairs. *a*, anal serrated hairs (*Aspidiotus nerii*); *b*, spiracular spines (*Ctenochiton*); *c*, marginal spines (*Ctenochiton*); *d*, lanceolate marginal spines (*Inglisia*); *e*, spines on anal tubercles (*Rhizococcus*); *f*, marginal spines (*Eriococcus*); *g*, conical spines (*Eriococcus*); *k*, marginal hairs (*Dactylopius*); *m*, hairs (*Icerya*); *n*, anal hairs (*Cœlostoma*).

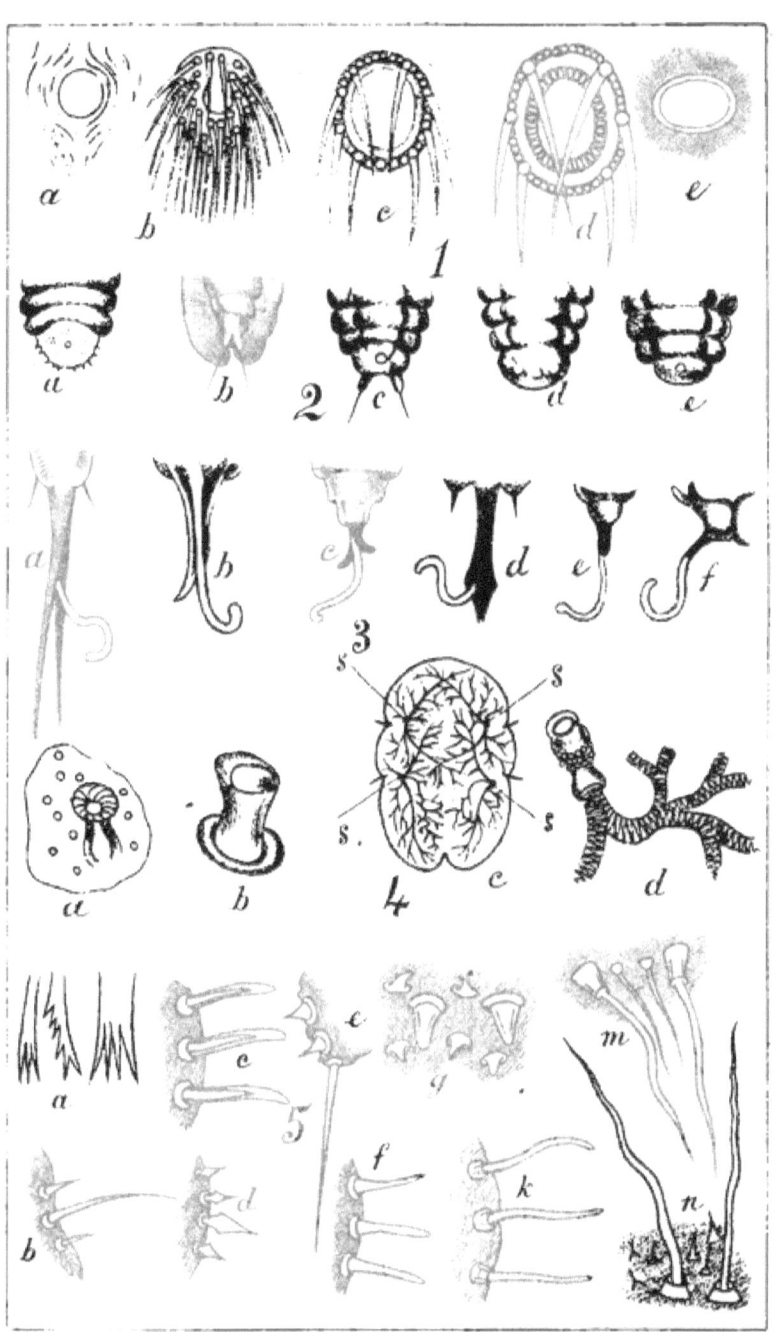

PLATE II.

Plate III.
Types of Last Abdominal Segments of the Diaspidinæ.

Fig.
1. *Aspidiotus coccineus* (after Comstock); no groups of spinnerets.
2. *Diaspis rosæ*; five groups of spinnerets.
3. *Mytilaspis cordylinidis*; five groups of spinnerets and single orifices.
4. *Mytilaspis pyriformis*; five groups of spinnerets, almost forming an arch.
5. *Chionaspis dysoxyli*; five groups of spinnerets.
6. *Poliaspis media*; eight groups of spinnerets.
7. *Fiorinia asteliæ*; arch of spinnerets.
8. *Fiorinia stricta*; five groups of spinnerets, the three upper groups almost forming an arch.

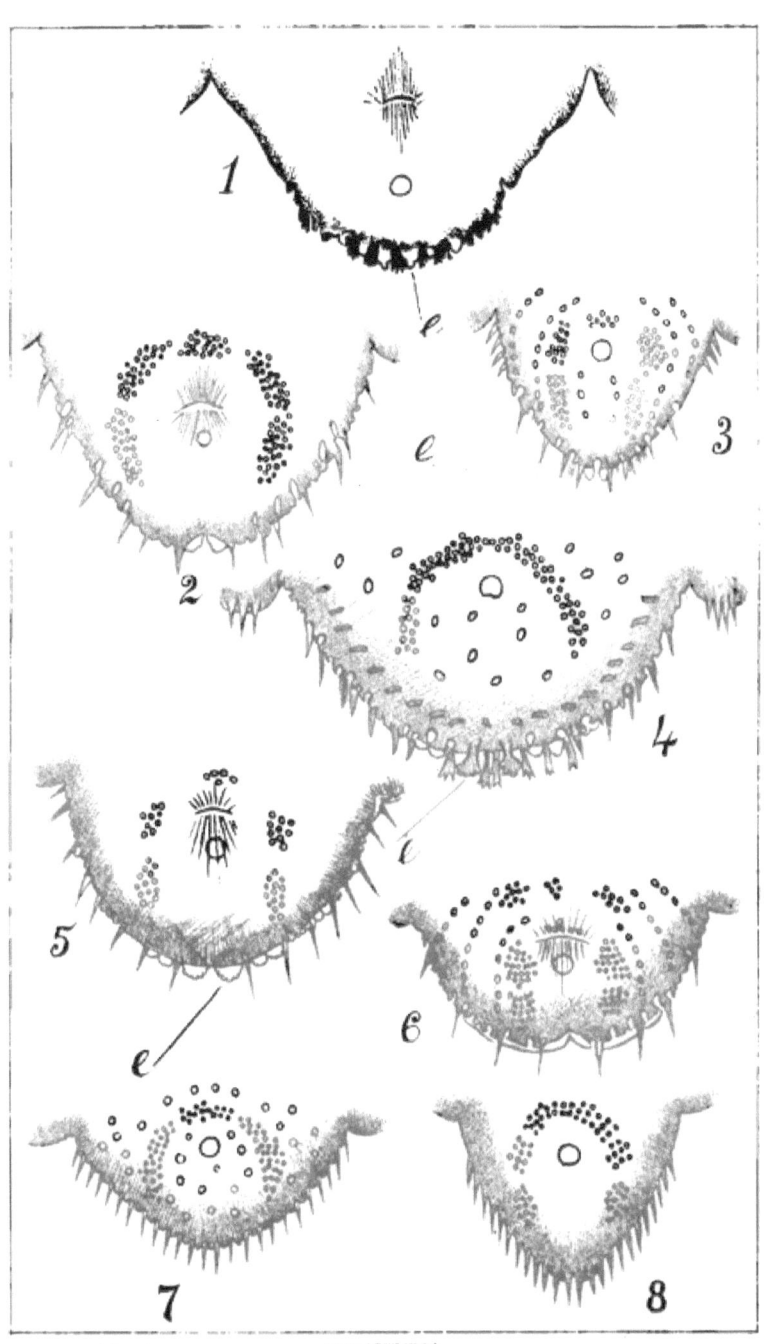

PLATE III.

Plate IV.

Fig.
1. *Aspidiotus atherospermæ.* a, insects on leaf of Atherosperma; b, adult female.

2. *Aspidiotus camelliæ.* a, insects on twig of Euonymus; b, adult female; c, d, puparia of female.

3. *Aspidiotus coccineus.* a, insects on rind of orange; b, adult female in puparium (overturned); c, puparium of male.

4. *Aspidiotus nerii.* a, insects on leaf of wattle; b, puparia, male and female.

5. *Diaspis Boisduvalii*, adult female (after Signoret).

6. *Diaspis rosæ.* a, insects on twig of rose; b, adult female; c, puparia, male and female.

7. *Diaspis santali.* a, insects on twig of pear; b, male and female puparia; c, last abdominal segment of female.

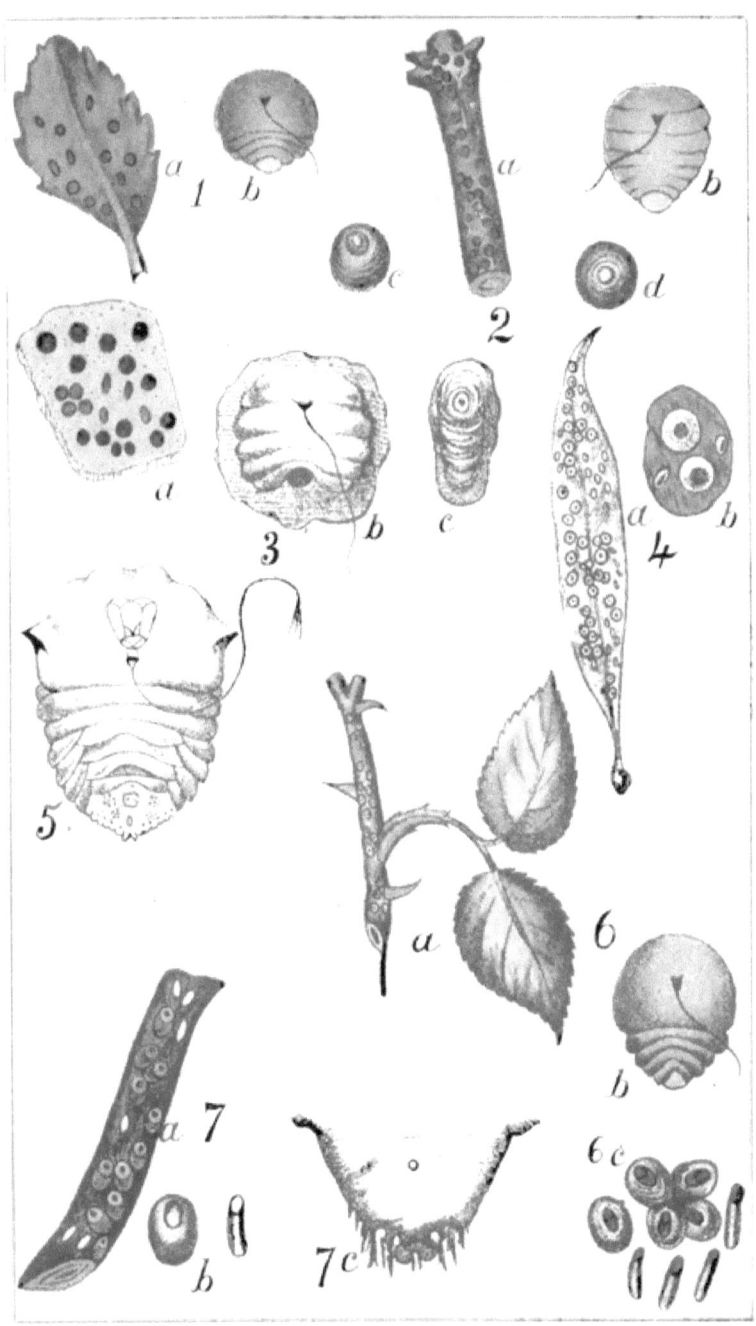

PLATE IV.

Plate V.

Fig.
1. *Mytilaspis cordylinidis.* a, insects on leaf of Cordyline australis (cabbage-tree); b, male and female puparia; c, adult female.

2. *Mytilaspis epiphytidis.* a, female puparium; b, male puparium; c, adult female.

3. *Mytilaspis drimydis.* a, insects on leaf of Drimys colorata; b, male and female puparia; c, adult female; d, marginal spinnerets.

4. *Mytilaspis leptospermi.* a, insects on bark of Leptospermum (manuka); b, male and female puparia; c, adult female.

5. *Mytilaspis pomorum.* a, insects on twig of hawthorn; b, female puparia; c, adult female; d, puparium overturned, showing enclosed female, f, and eggs, e.

6. *Mytilaspis pyriformis.* a, insects on leaf of Dysoxylon spectabile; b, male and female puparia; c, adult female; d, male; e, last three joints of male antenna; f, foot of male.

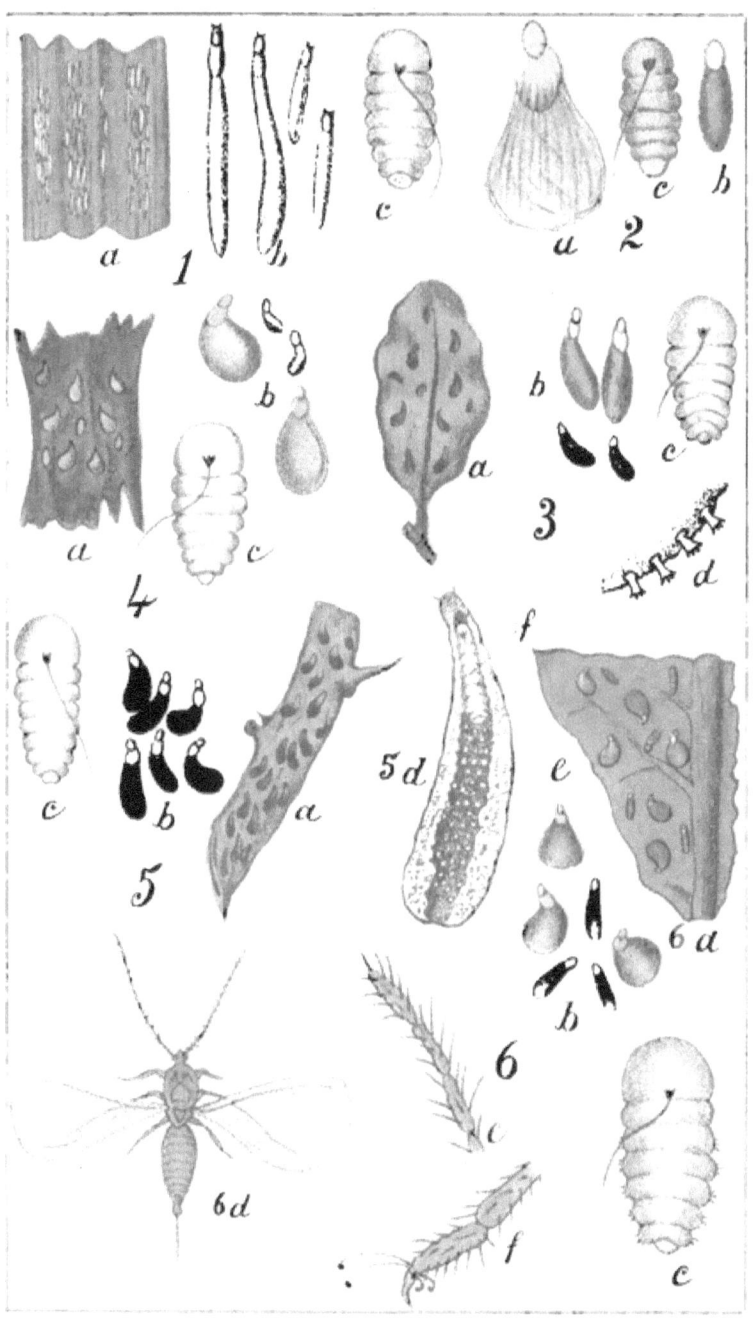

PLATE V.

PLATE VI.

Fig.
1. *Chionaspis citri.* a, insects on rind of orange; b, male and female puparia c, adult female.

2. *Chionaspis dubia.* a, insects on fern-leaf (Pellæa); b, male and female puparia; c, adult female.

3. *Chionaspis dysoxyli.* a, insects on leaf of Dysoxylon; b, male and female puparia; c, adult female.

4. *Chionaspis minor.* a, insects on twig of Parsonsia; b, male and female puparia; c, adult female.

5. *Poliaspis media.* a, insects on leaves of Cyathodes; b, male and female puparia; c, adult female.

6. *Fiorinia asteliæ.* a, insects on leaves of Coprosma; b, male and female puparia; c, adult female; d, pellicles of second stage.

7. *Fiorinia stricta.* a, insects on leaves of Earina; b, male and female puparia; c, adult female.

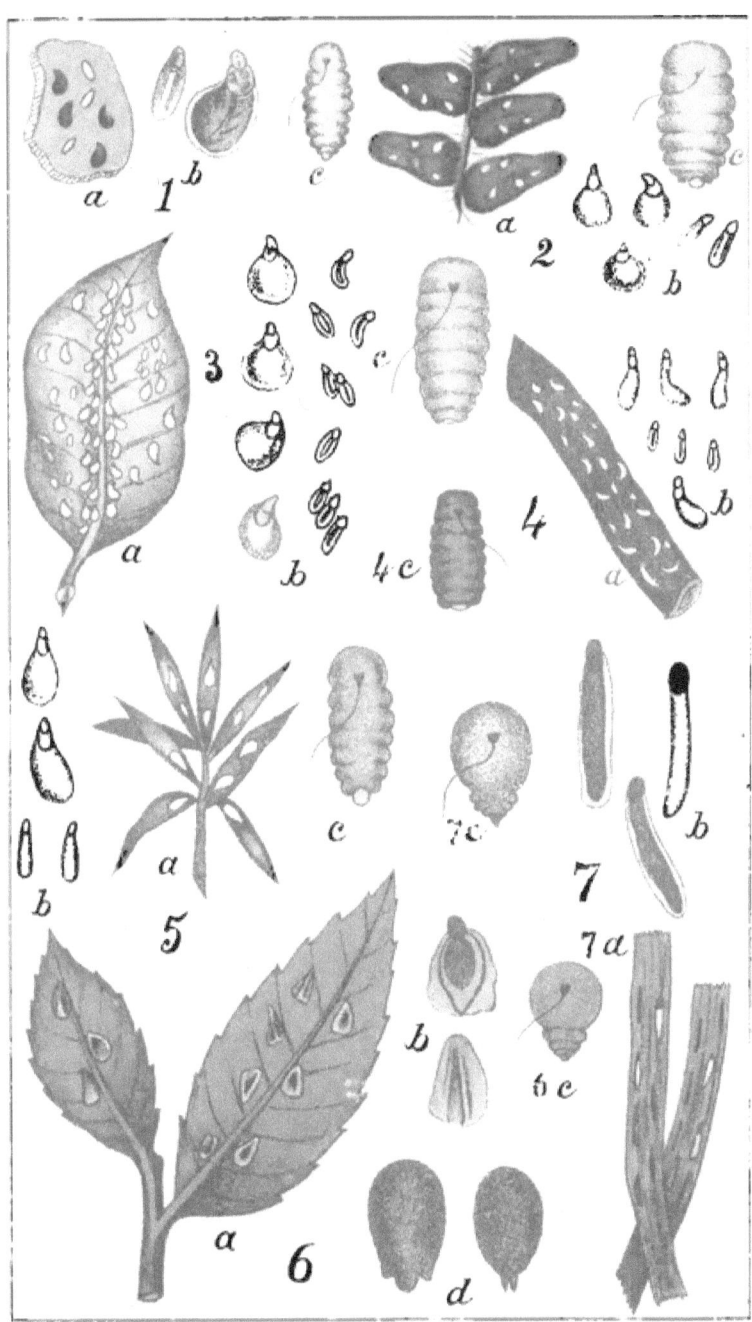

PLATE VI.

PLATE VII.

Fig.
1. *Lecanochiton metrosideri.* a, insects on branch and leaves of Metrosideros (rata); b, test of female, upper side; c, test of female, under-side; d, female of second stage, ventral aspect; e, antenna of adult female; f, test of male.

2. *Ctenochiton depressus.* a, insects on leaf of Plagianthus; b, adult female, dorsal aspect; c, test of male; d, part of fringe of female test.

3. *Ctenochiton elæocarpi.* a, insect on bark of Elæocarpus (hinau); b, female of second stage in test, on leaf.

4. *Ctenochiton elongatus.* a, female tests on leaves of Earina; b, part of fringe.

5. *Ctenochiton flavus.* a, insects on half leaf of Panax; b, female tests; c, male test; d, adult female, ventral aspect; e, spiracular and marginal spines and spinnerets of female.

6. *Ctenochiton fuscus.* a, insects on half leaf of Panax; b, female test, upper side.

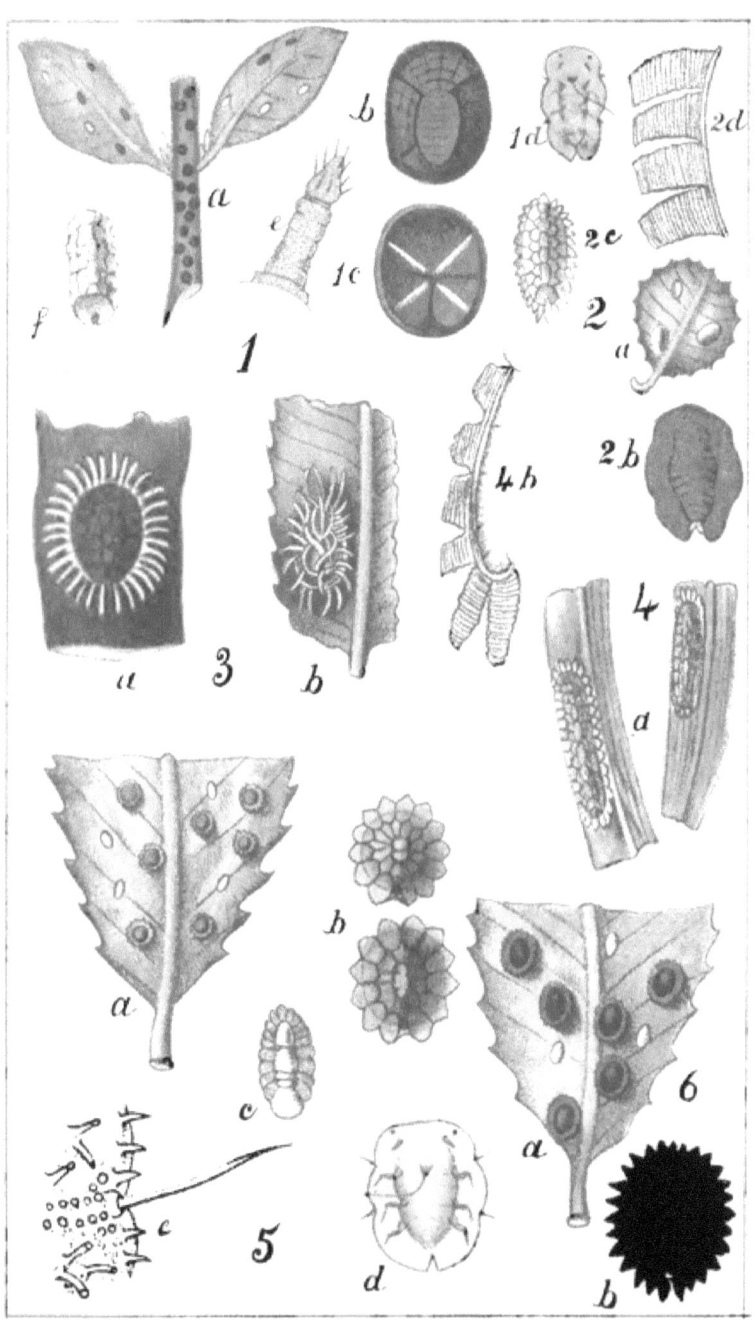

PLATE VII.

PLATE VIII.

Fig.
1. *Ctenochiton hymenantheræ.* a, insects on leaves of Hymenanthera; b, female test; c, male test.

2. *Ctenochiton perforatus.* a, insects on leaf of Coprosma; b, adult female in test, ventral aspect; c, male test; d, female test and fringe; e, portion of edge of female test with three segments of fringe, showing the rows of perforations; f, antenna of adult female; g, male; k, head of male, upper side; m, head of male, under-side.

3. *Ctenochiton piperis.* a, insects on leaf of *Piper excelsum;* b, females in tests, dorsal aspect; c, male pupa in test, dorsal aspect; d, female of second stage, dorsal aspect; e, two of the tubercles on the dorsum of the adult female, after pressure.

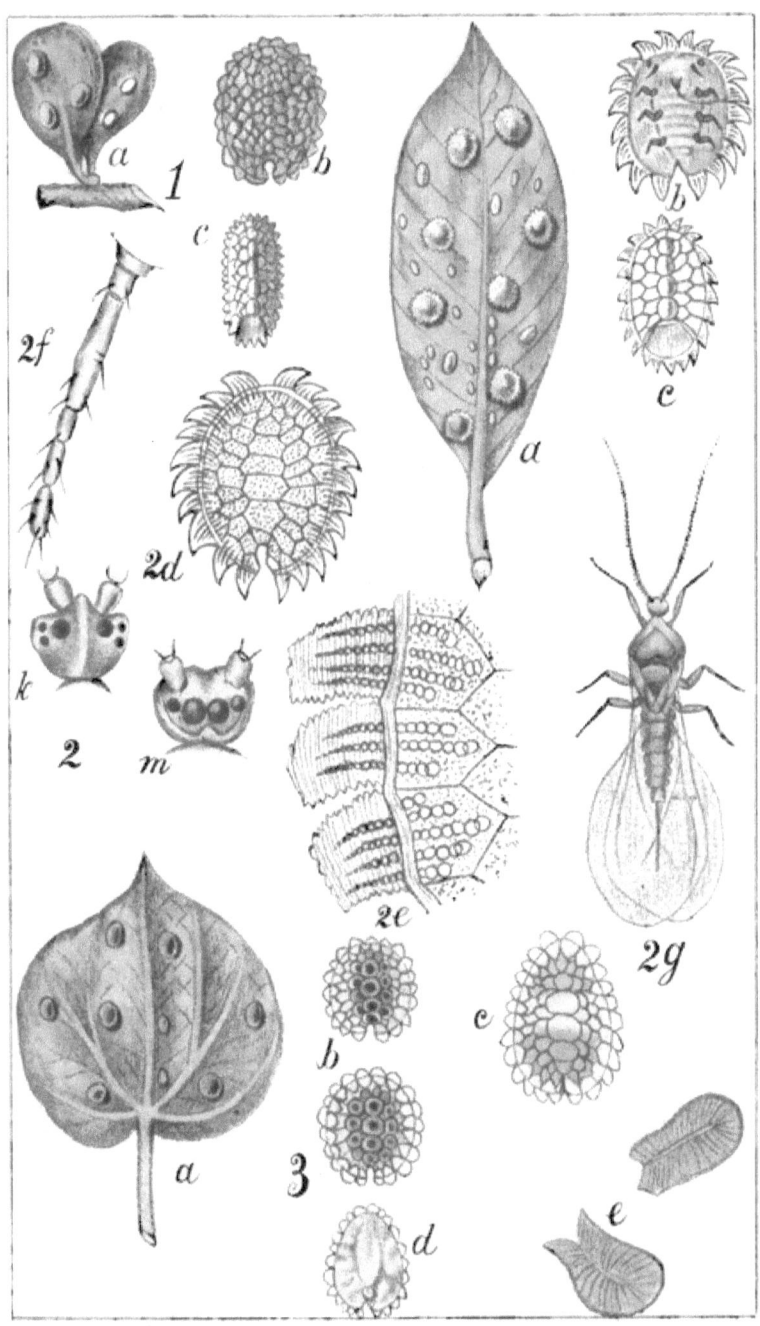

PLATE VIII.

PLATE IX.

Fig.
1. *Ctenochiton viridis.* a, insects on leaf of Coprosma; b, female of second stage; c, adult female (the test removed); d, male test; e, a segment of the female test; f, diagram of head of male, upper and under sides shown together, eight eyes and two ocelli; k, antenna of male.

2. *Inglisia leptospermi.* a, insects on twig of Leptospermum (manuka); b, female test, side view; c, female test, upper side; d, male test; e, a segment of the female test; f, adult female, dorsal aspect; g, antenna of female; k, foot of female.

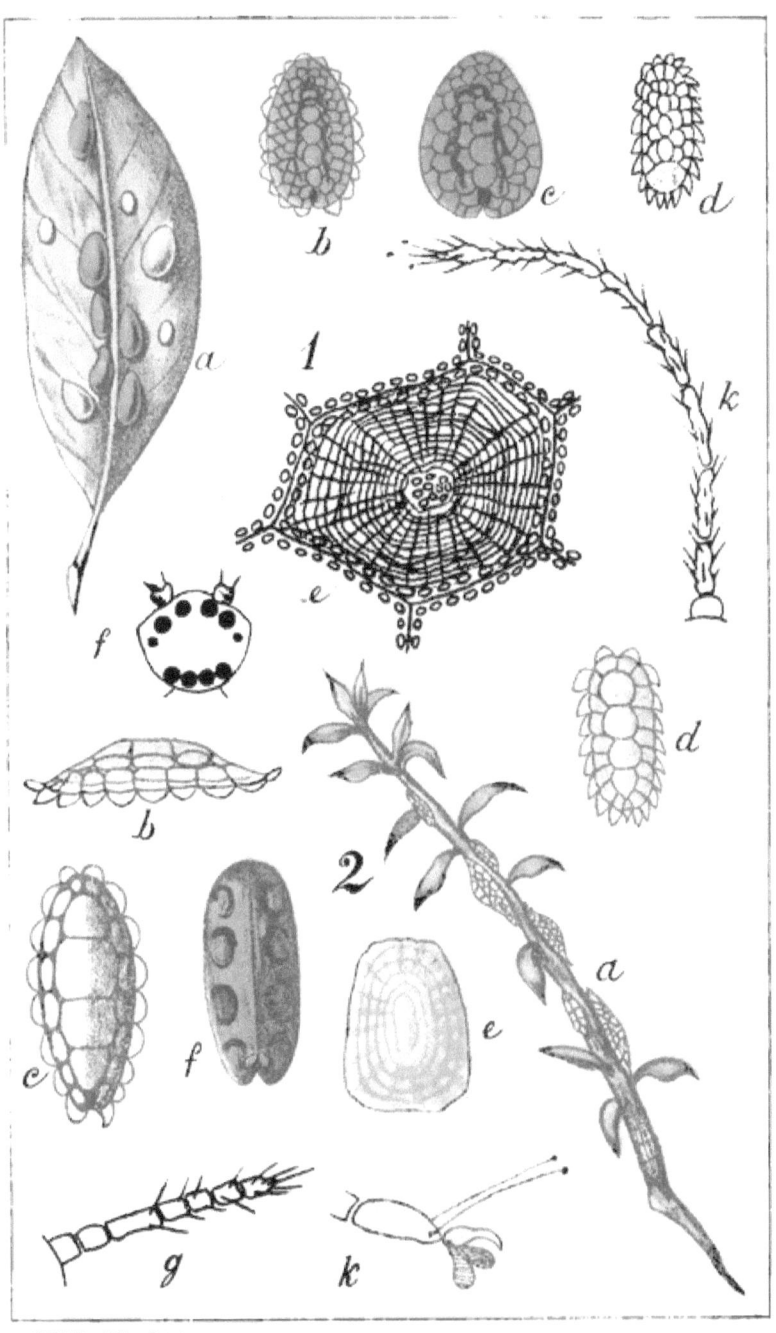

PLATE IX.

PLATE X.

Fig.
1. *Inglisia ornata.* a, insects on twig of Elæocarpus (hinau); b, female test, side view; c, female test, dorsal view; d, a segment of the female test and three segments of fringe; e, male test; f, adult female, side view; g, marginal spines of female.

2. *Inglisia patella.* a, insects on leaf of Coprosma; b, c, female tests, upper side; d, adult female, ventral aspect; e, marginal spines of female.

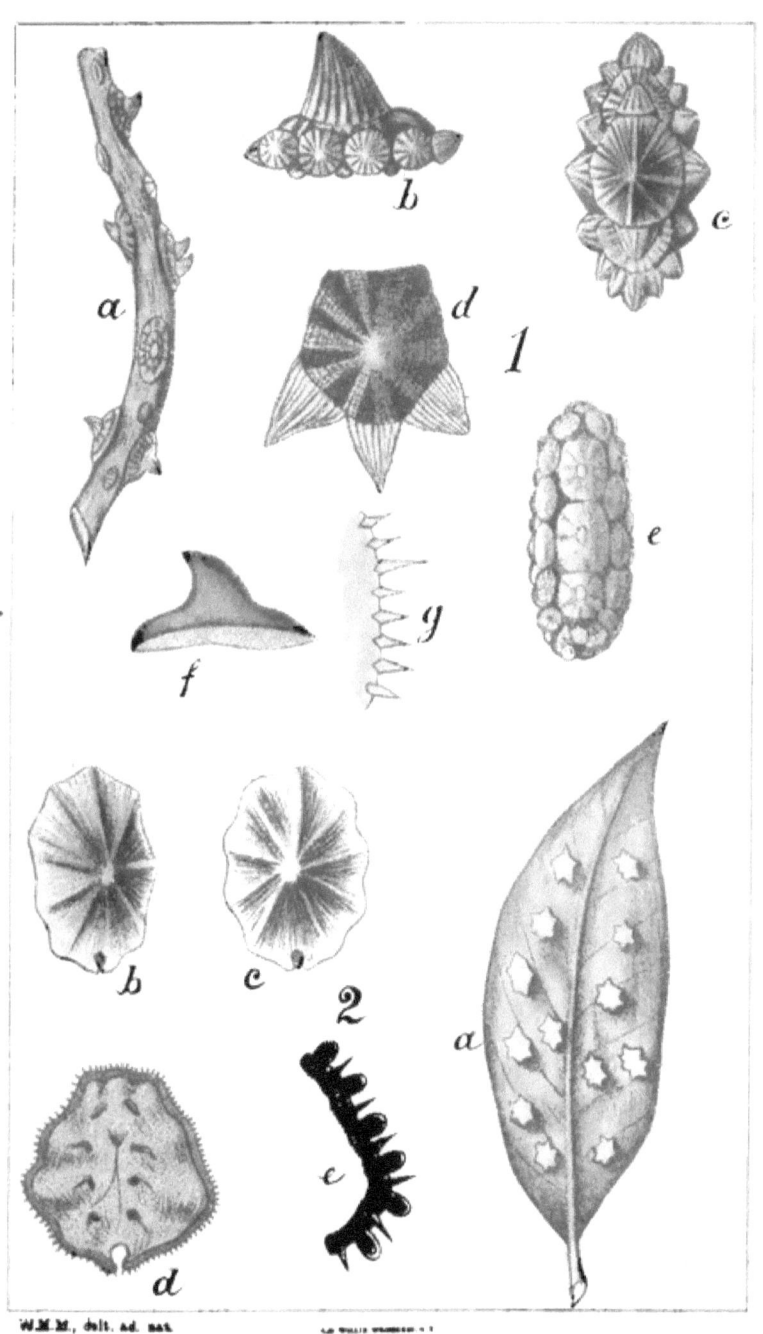

PLATE X.

PLATE XI.

Fig.
1. *Lecanium depressum.* a, adult female, dorsal aspect; b, markings of the skin.

2. *Lecanium hemisphæricum.* a, insects on Camellia; b, adult female, dorsal aspect; c, adult female, side view; d, female of second stage.

3. *Lecanium hesperidum.* a, insects on leaf of ivy; b, adult female, dorsal aspect; c, female of second stage; d, markings of skin.

4. *Lecanium mori.* a, insects on fern-leaf (Asplenium); b, adult female, dorsal aspect.

5. *Lecanium oleæ.* a, insects on twig of Camellia; b, adult female, dorsal aspect; c, adult female, side view; d, female of second stage, dorsal aspect; e, markings of skin.

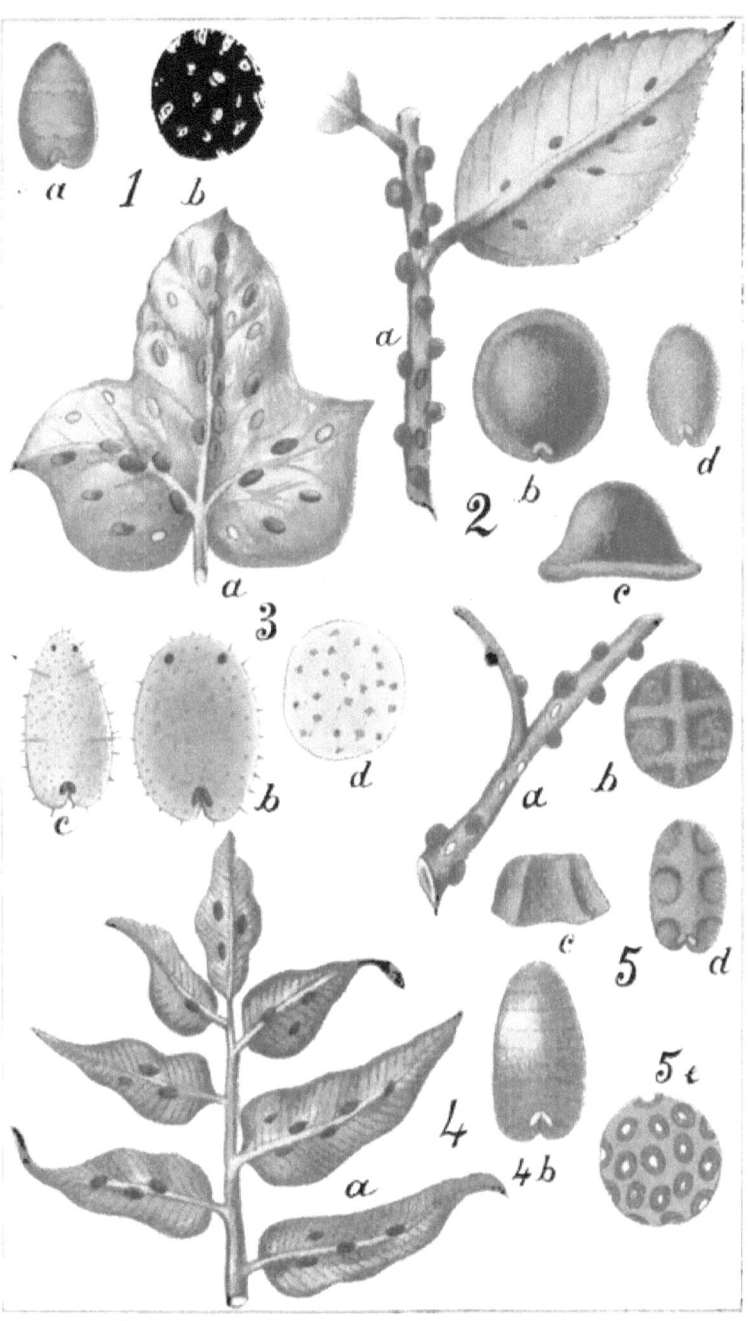

PLATE XI.

Plate XII.

Fig.
1. *Pulvinaria camellicola.* a, insects on branch and leaf of Camellia; b, adult female and ovisac, dorsal aspect; c, adult female and ovisac, side view; d, adult female, dorsal aspect; e, female of second stage; f, markings of skin, with hairs; g, antenna of adult female; h, diagram of head of male, upper and under sides shown together, four eyes and two ocelli.

2. *Planchonia epacridis.* a, insect in test on leaf of Leucopogon; b, female in test, dorsal aspect; c, portion of the double fringes; d, extremity of abdomen of female; e, anogenital ring and anal tubercles of female; f, antenna of larva; g, rings with hairs replacing antennæ of adult female.

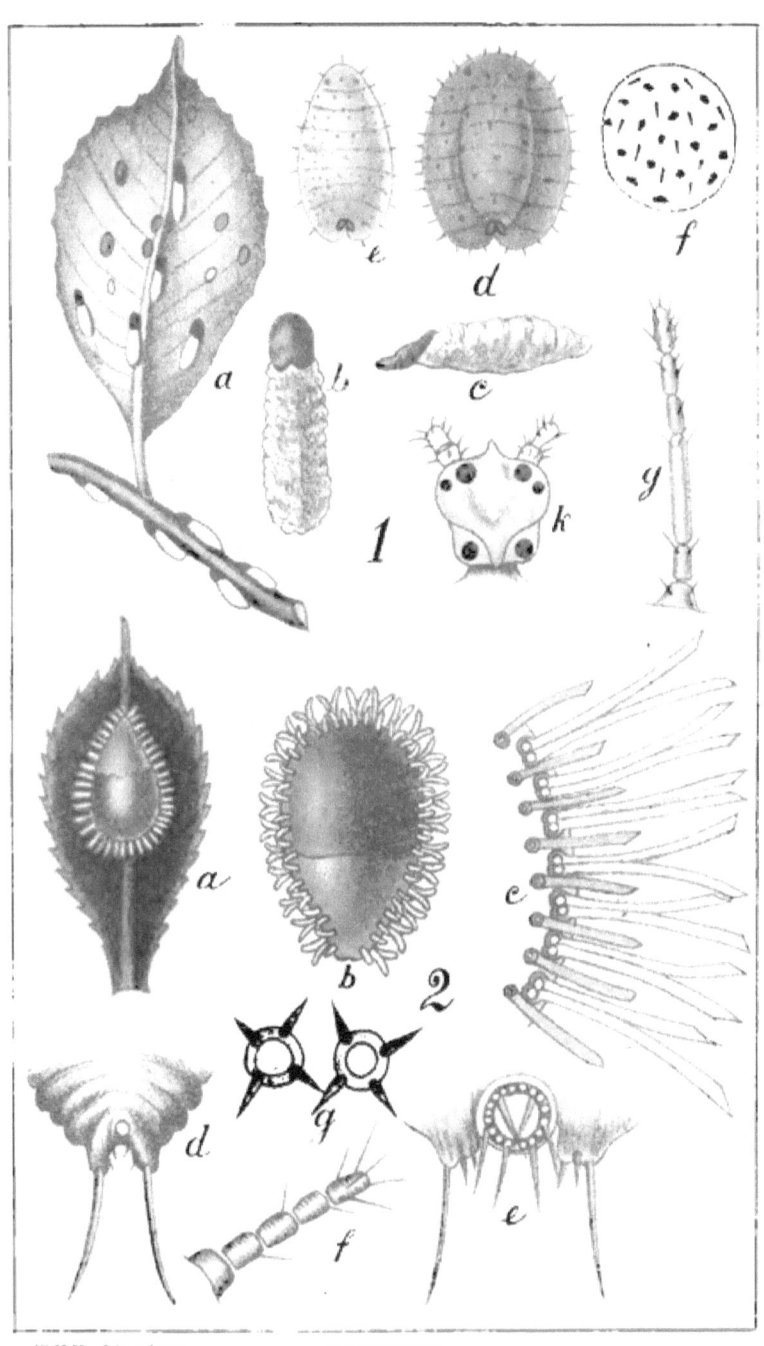

PLATE XII.

PLATE XIII.

Fig.
1. *Eriochiton hispidus.* a, Insects on twig and leaves of Olearia Haastii; b, adult female, dorsal aspect, with fragments of test and fringe; c, adult female without test, dorsal aspect; d, male test, upper side; e, male test, under-side; f, larva, with fringe; g, female of second stage, without test, dorsal aspect; k, m, spines and tubular fringe; n, antenna of adult female; p, foot of adult female; s, last five joints of male antenna; t, abdominal spike of male.

2. *Eriochiton spinosus.* a, insects on twig of Melicope; b, adult female, without test; c, female of second stage; d, male test; e, marginal spines and feathery fringe of female; f, foot of adult female; g, antenna of adult female.

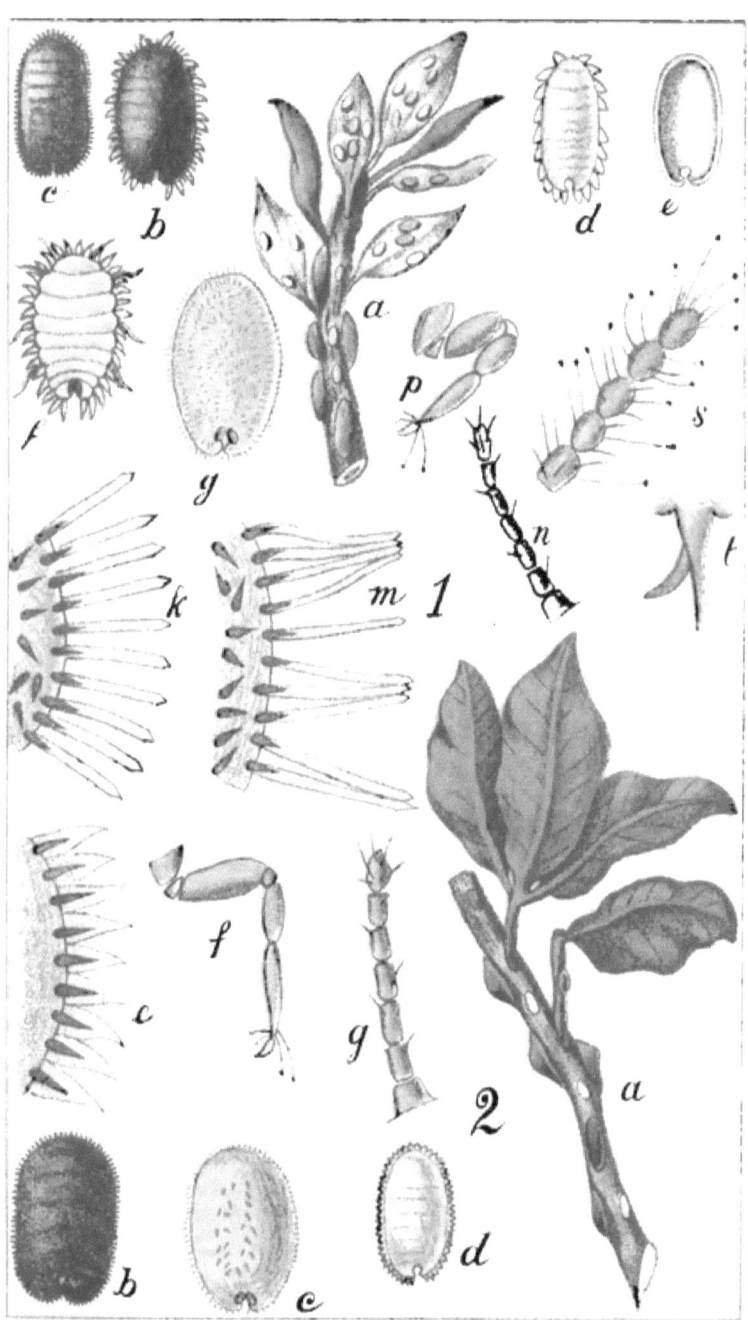

PLATE XIII.

PLATE XIV.

Fig.
1. *Eriococcus araucariæ.* a, insects on twig of Araucaria excelsa (Norfolk Island pine); b, sac of female, upper side; c, sac of female, under-side, with enclosed shrivelled female and eggs; d, sac of male; e, adult female before gestation; f, extremity of abdomen, anogenital ring, and anal tubercles of female.

2. *Eriococcus hoheriæ.* a, insects on bark of Hoheria; b, sac of female, upper side; c, sac of male; d, adult female; e, extremity of abdomen and anal tubercles of adult female; f, foot of adult female; g, antenna of adult female; k, larva, ventral aspect; m, male; n, abdominal spike of male: p, antenna of male.

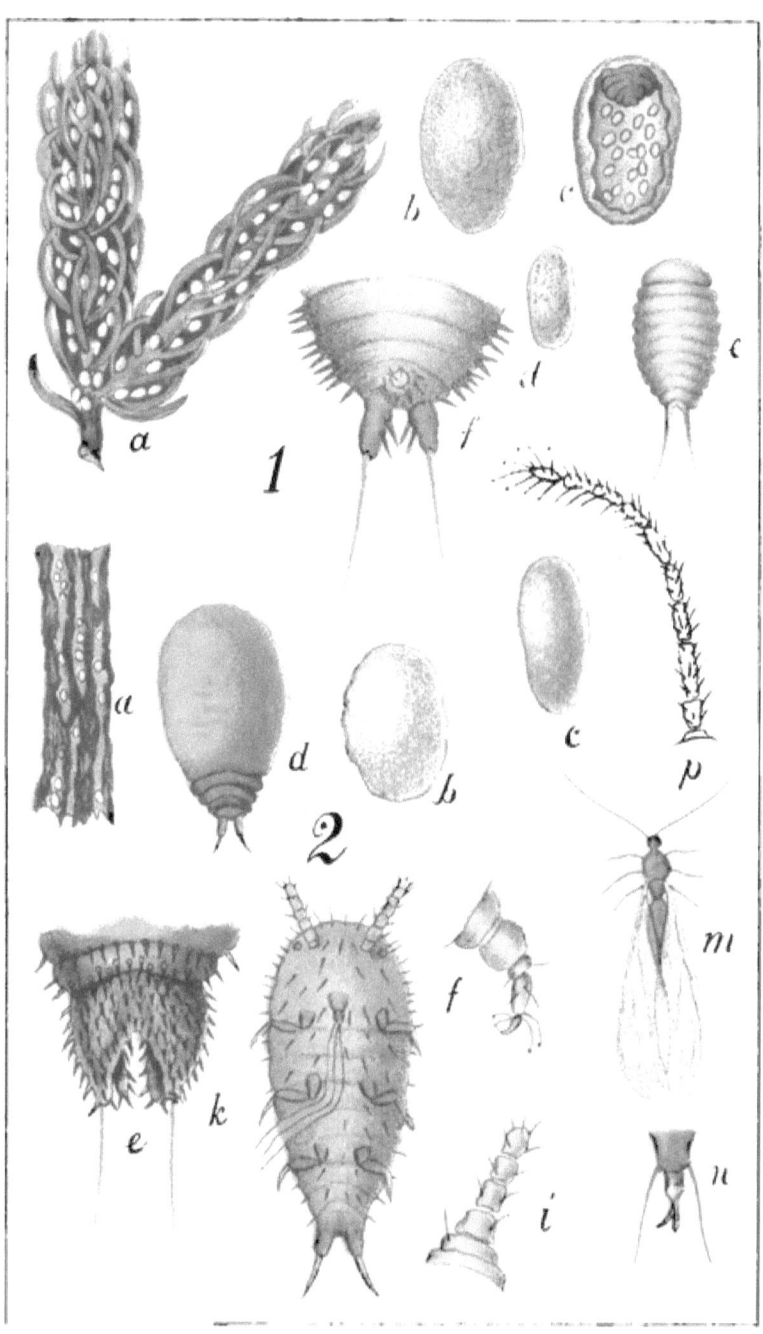

PLATE XIV.

PLATE XV.

Fig.
1. *Eriococcus multispinus.* a, insects on leaf of Knightia; b, sac of female, upper side; c, sac of female, under-side, with enclosed insect; d, adult female, dorsal aspect; e, diagram of arrangement of spines on female; f, spines of female; i, antenna of female; k, head of male, upper side; m, abdominal spike of male.

2. *Eriococcus pallidus.* a, insects on leaf of Myoporum (ngaio); b, sac of female; c, adult female, dorsal aspect; d, part of abdomen of female; e, antenna of female.

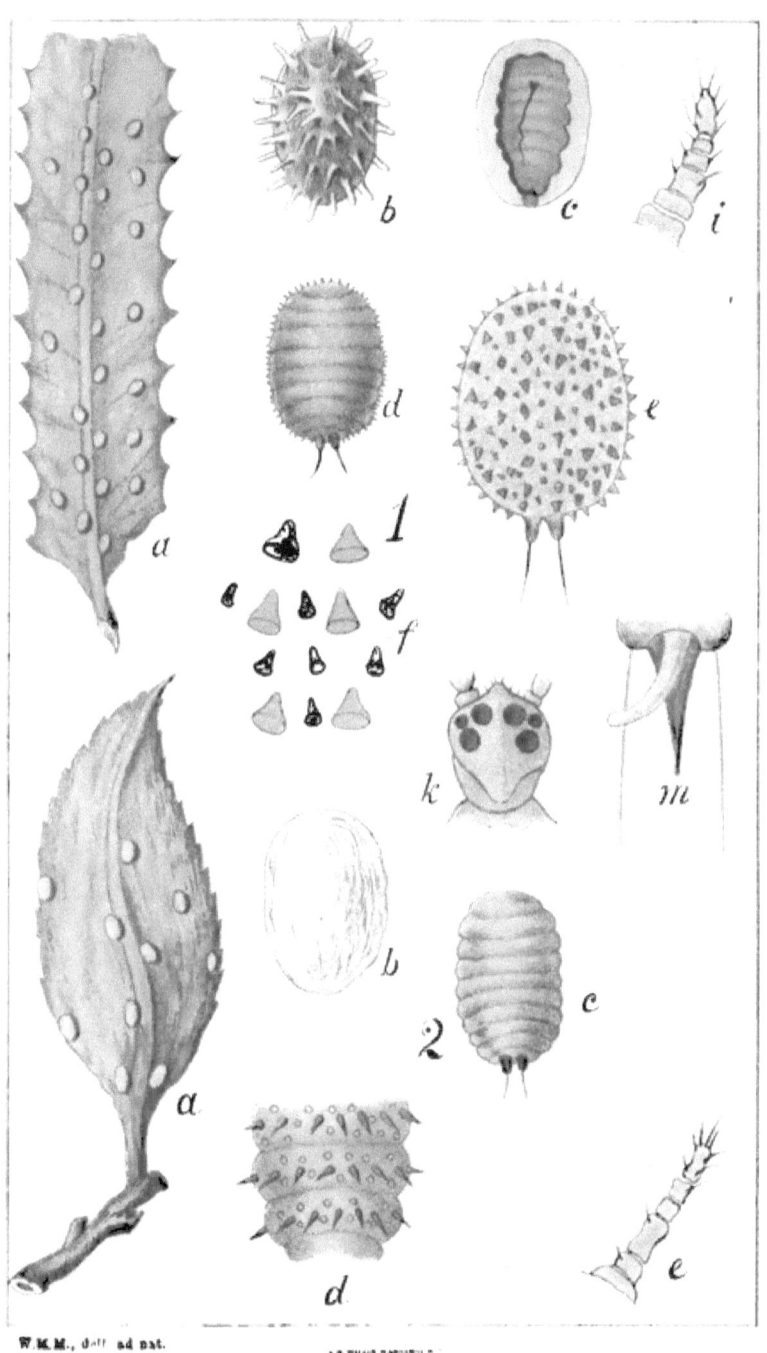

PLATE XV.

Plate XVI.

Fig.
1. *Rhizococcus celmisiæ.* a, Insects on leaves of Celmisia; b, adult female dorsal aspect.

2. *Rhizococcus fossor.* a, Insects on leaves of Santalum (maire); b, elevation produced by insect on *upper* side of leaf; c, female in pit on *under*-side of leaf; d, sac of male; e, adult female before gestation, from a pit; f, adult female from surface of leaf; g, female of second stage; k, marginal spines and tubular cotton of second stage; m, antenna of adult female; n, male; p, diagram of head of male, upper and lower sides shown together, four eyes and two ocelli; s, abdominal spike of male; t, antenna of male.

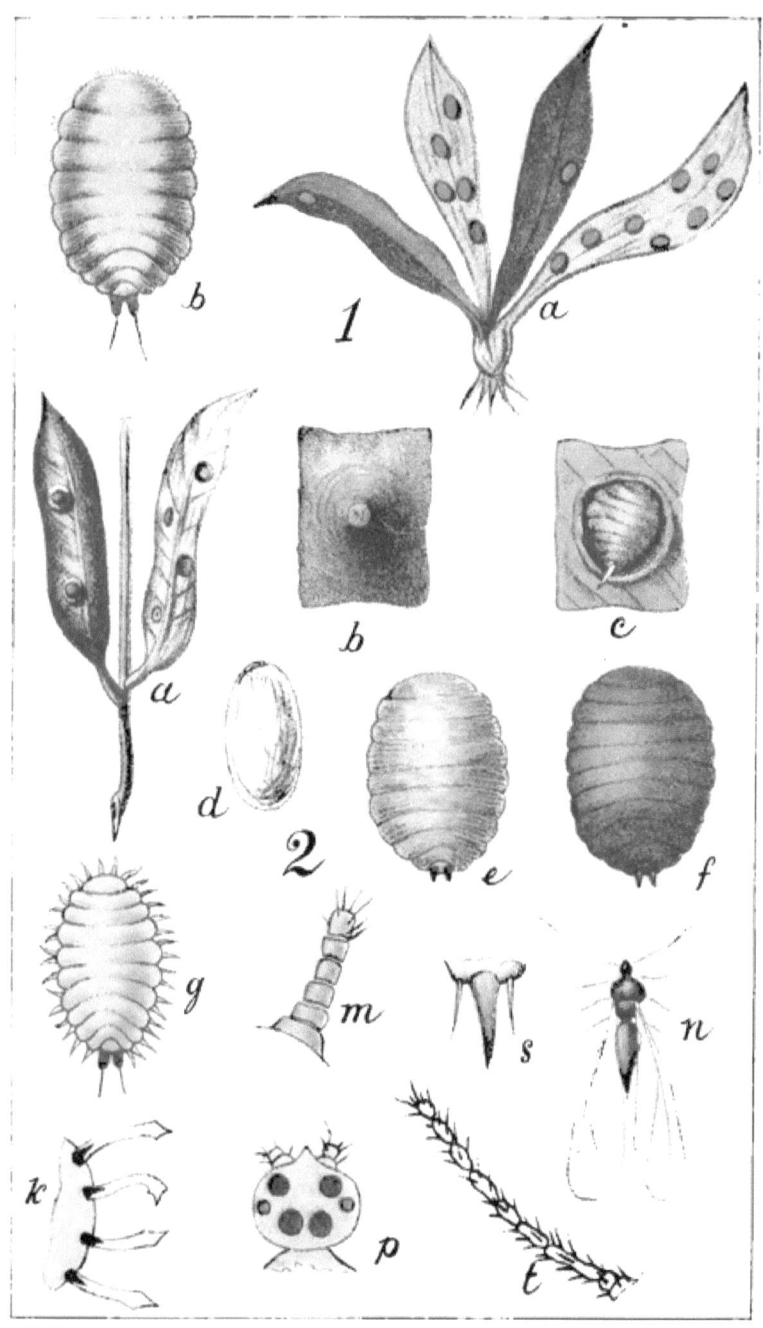

PLATE XVI.

Plate XVII.

Fig.
1. *Dactylopius alpinus.* a, Insects on twig of Veronica; b, adult female, dorsal aspect; c, female of second stage; d, anogenital ring; e, antenna of adult female.

2. *Dactylopius calceolariæ.* a, Insects on leaf of Phormium tenax; b, adult female, dorsal aspect; c, antenna of adult female.

3. *Dactylopius glaucus.* a, Insects on leaf of Coprosma; b, adult female, green variety; c, adult female, brown variety; d, sac of male; e, antenna of male.

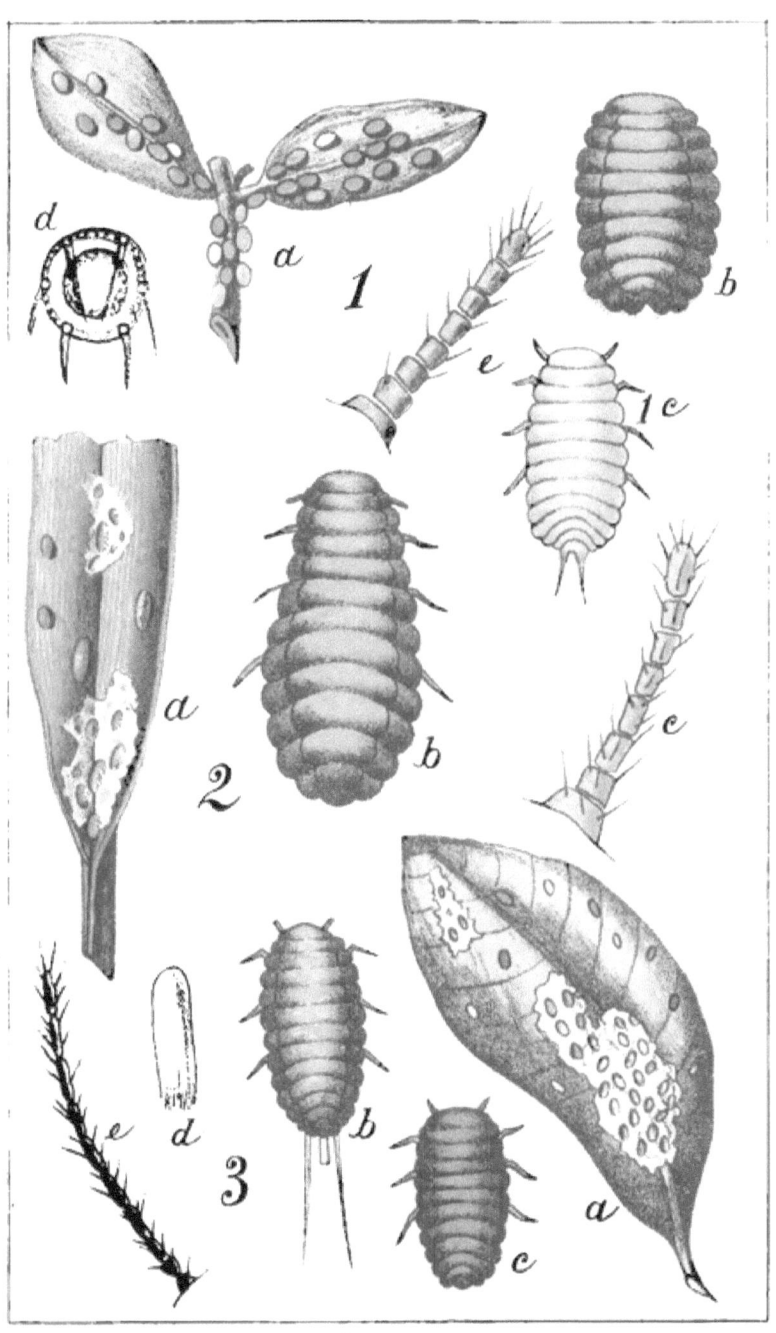

PLATE XVII.

PLATE XVIII.

Fig.
1. *Dactylopius poæ.* a, Insects on roots of Poa (tussock-grass); b, adult female; c, antenna of adult female; d, foot of adult female; e, anogenital ring and anal tubercles of adult female.

2. *Pseudococcus asteliæ.* a, Insects on leaf of Astelia; b, adult female, dorsal aspect; c, antenna of adult female; d, foot of adult female; e, various spinnerets.

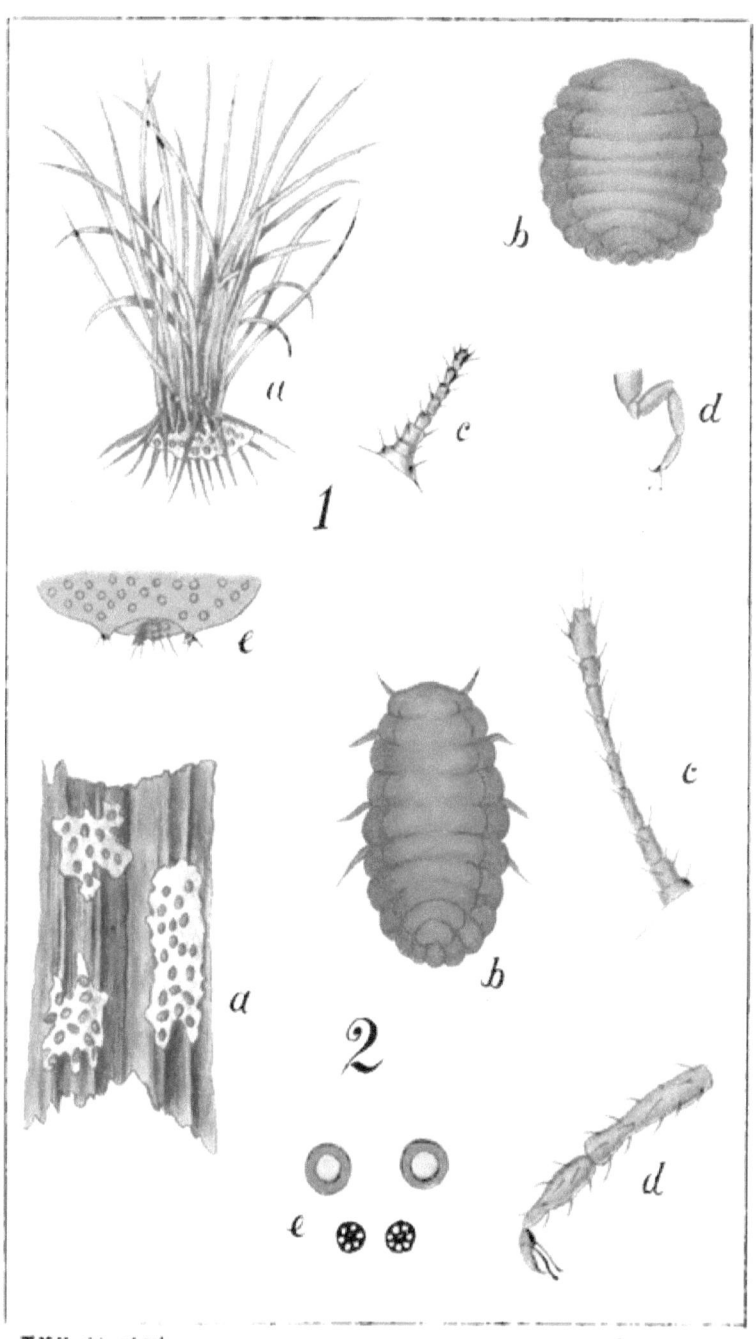

PLATE XVIII.

PLATE XIX.

Icerya Purchasi. *a*, Insects on twig of Acacia (wattle); *b*, adult female and ovisac, upper view; *c*, adult female and ovisac, side view; *d*, female of second stage; *e*, larva, with yellow cotton; *f*, male; *g*, haltere of male *k*, two joints of male antenna; *m*, hairs, spinnerets, and glassy tubes *n*, antenna of adult female.

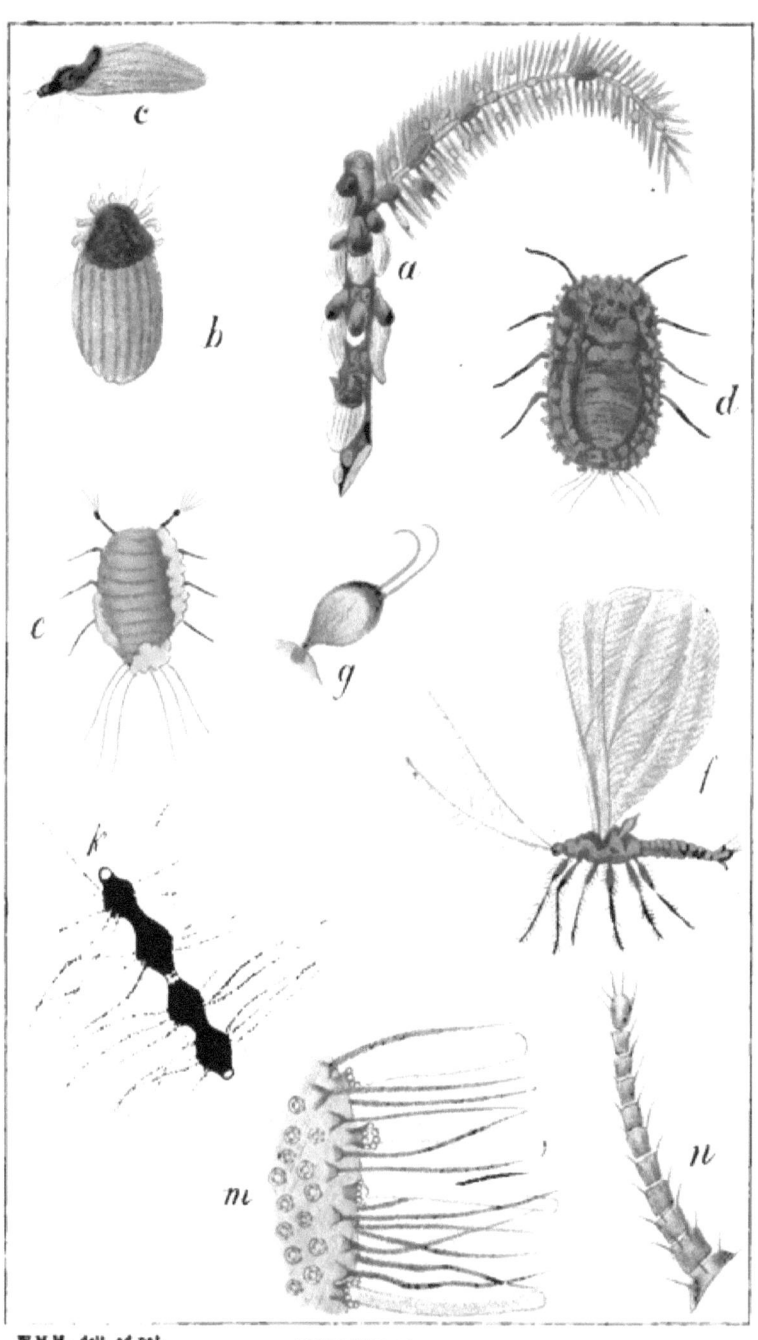

PLATE XIX.

PLATE XX.

Cœlostoma zelandicum. a, adult female, dorsal aspect; b, adult female, ventral aspect; c, waxy tests of females of second stage on twig of Muhlenbeckia; d, female of second stage, dorsal aspect; e, female of second stage, ventral aspect, with anal tuft of cotton and seta; f, larva, with yellow mealy secretion; g, eggs, in cottony mass; k, antenna of female, second stage; l, foot of female, second stage; m, antenna of adult female; n, spiracle and trachea of female; p, anal extremity and "honey-dew" organ of larva and female of second stage.

N.B.—The foot of the adult female is shown in Plate I., Fig. 6.

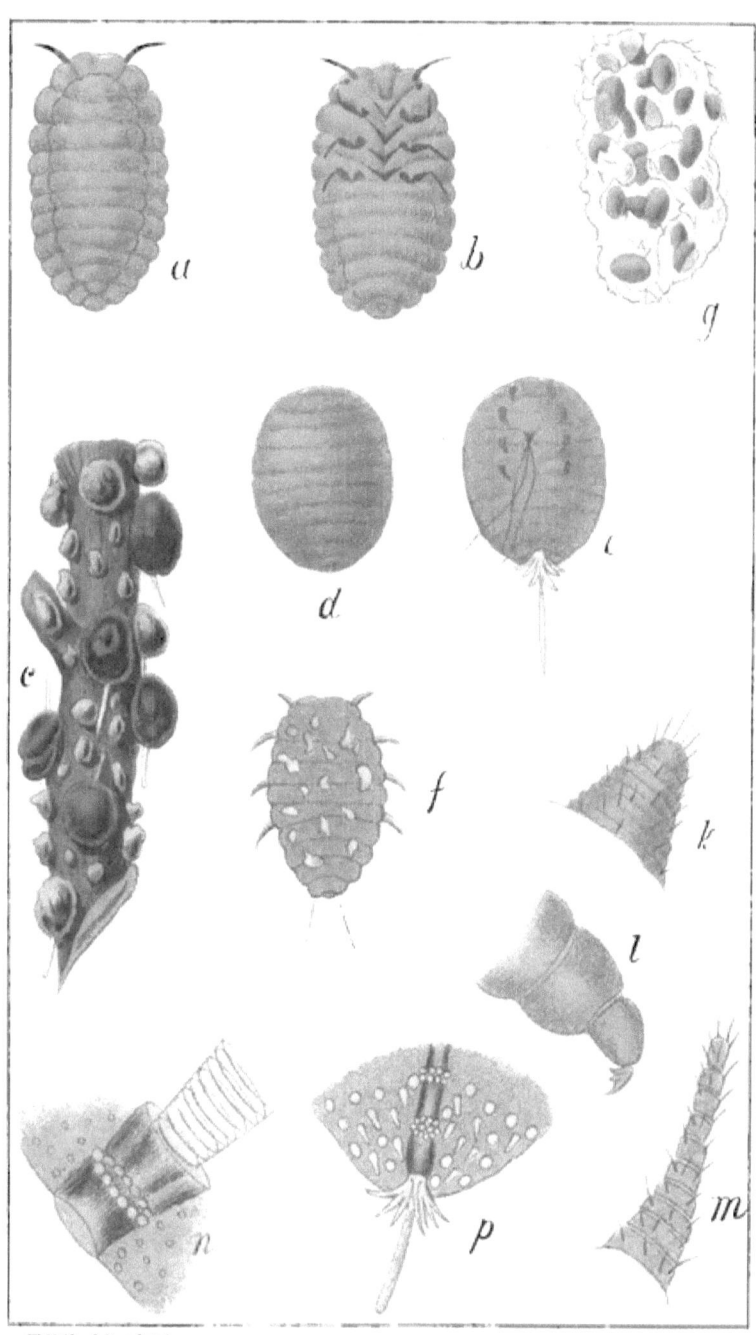

PLATE XX.

Plate XXI.

Fig.
1. *Cœlostoma zœlandicum.* *a*, male insect; *b*, abdomen of male; *c*, foot of male; *d*, hairs and marks, abdomen of male; *e*, antenna of male; *f*, circular marks, abdomen of male; *k*, abdominal spike and penis of male; *m*, haltere of male.

2. *Cœlostoma wairoense.* *a*, male insect; *b*, head of male, upper side, with facetted eyes; *c*, foot of male; *d*, circular marks, abdomen of male.

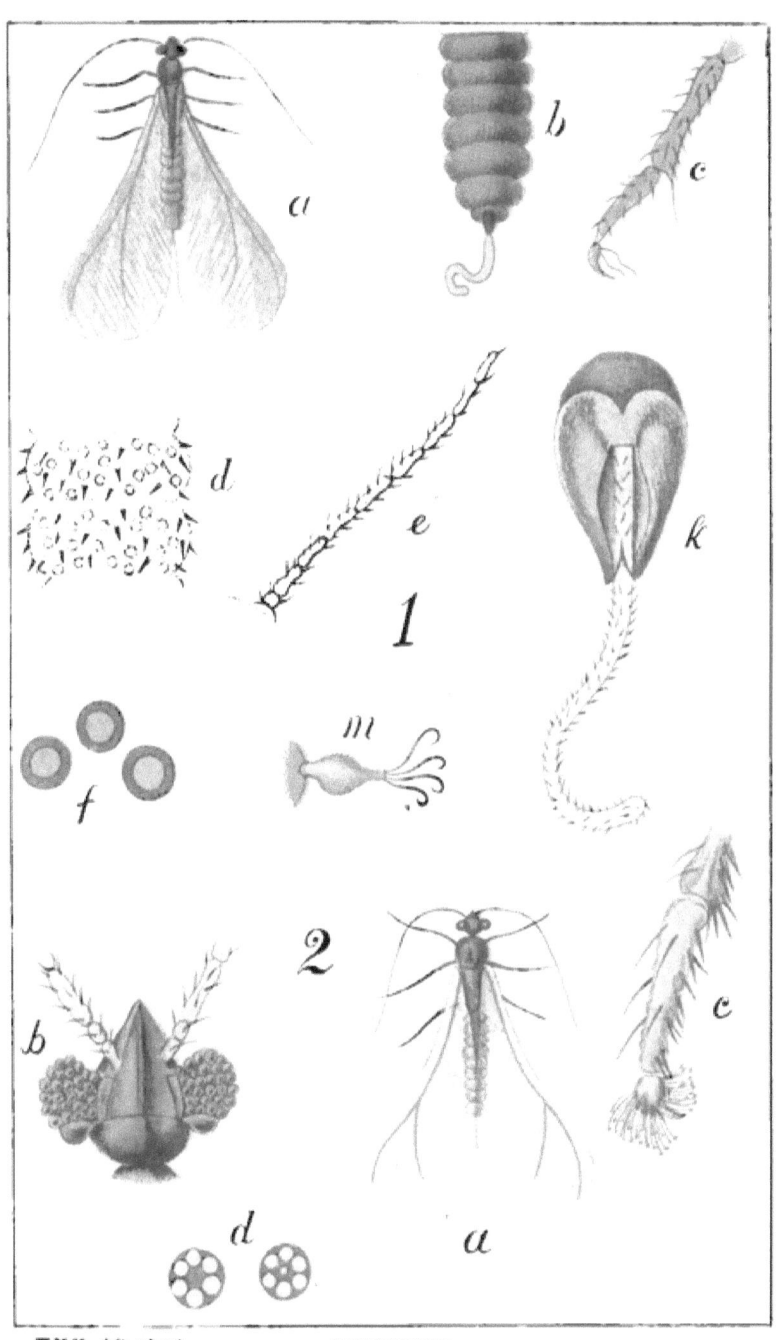

PLATE XXI.

Plate XXII.

The Honey-dew, and resulting Fungi.

Fig.
1. *a*, Lecanid female (Ctenochiton) with protruded honey-dew organ; *b*, abdominal extremity of the same, enlarged, the drop of honey-dew bursting in spray.

2. *a*, fungoid growth on upper side of leaf; *b*, fungoid growth on twig; *c, d, e,* various forms of black fungi from honey-dew.

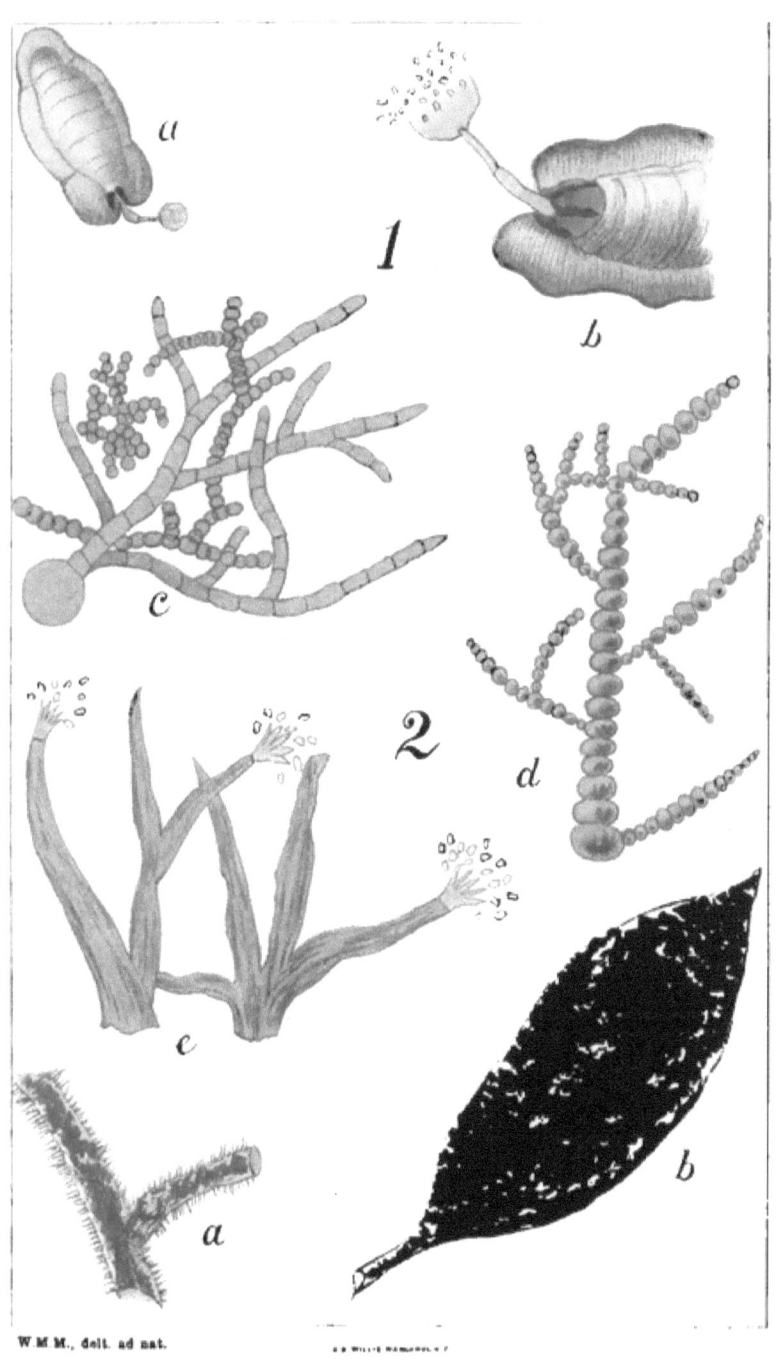

PLATE XXII.

Plate XXIII.
Parasites of Coccididæ.

Fig.
1. *a*, pupa of hymenopterous parasite; *b*, the same pupa under the waxy test of *Ctenochiton perforatus*; *c*, imago, or full-grown parasite.

2. *a*, brown and yellow fungi on *Ctenochiton viridis* (leaf of Hedycarya dentata); *b*, upper side of brown fungus; *c*, under-side of the same, with attached fungoid sheet; *d*. *Ctenochiton viridis* (test removed) filled with yellow fungus, and with globular mass of the same above it.

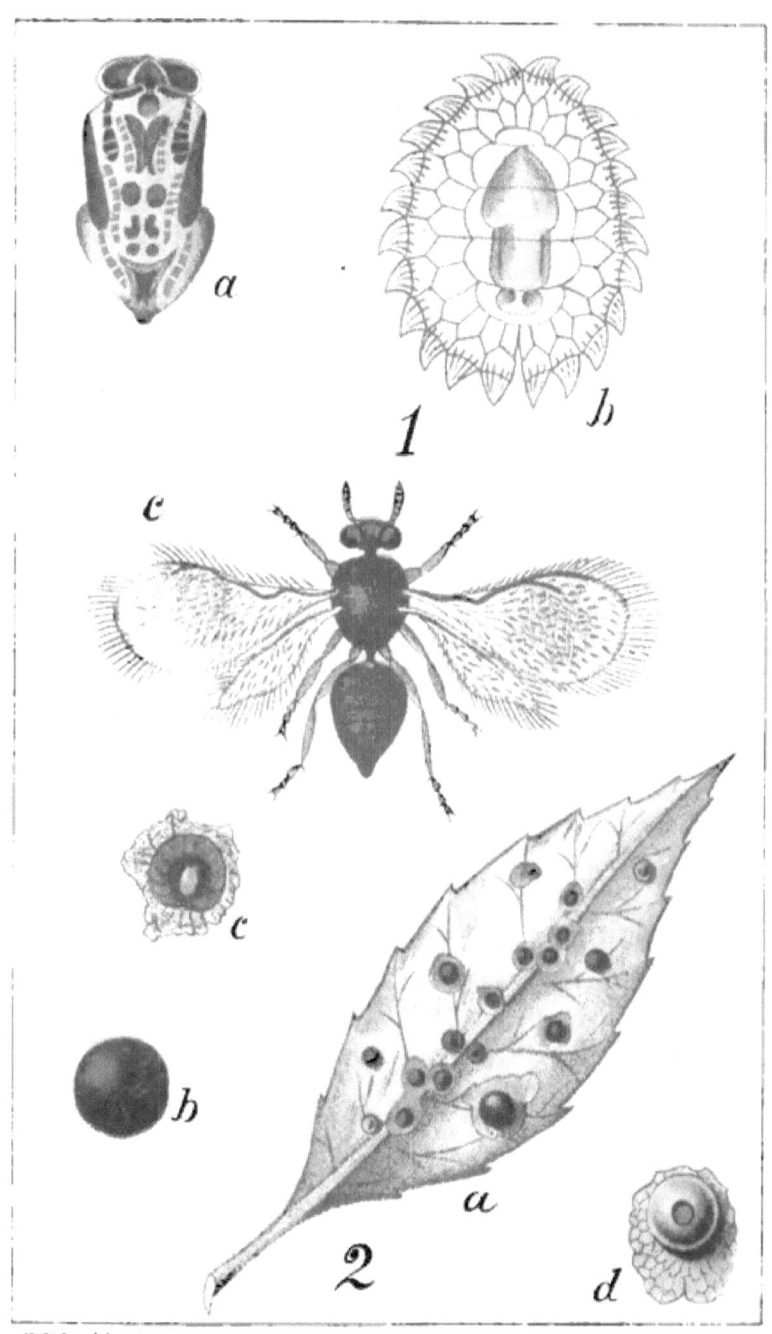

PLATE XXIII.

SINCE this work has been in type, the author has received a letter from the State Inspector of Fruit Pests for California, in which the writer states that the insect *Icerya Purchasi* has there, especially in the southern part of the State, gained such hold on the orange-groves, in spite of the most strenuous efforts, that the people find it impossible to keep it down. Orange- and lemon-growers (and indeed other tree-growers) in New Zealand, especially in the North Island, should take note of this fact, and beware of ever letting this omnivorous and most destructive insect obtain any footing on their trees. *A speedy burning of every infected tree is the best remedy in this case.*

www.ingramcontent.com/pod-product-compliance
Lightning Source LLC
Chambersburg PA
CBHW031827230426
43669CB00009B/1256